Secure, Low-Power IoT Communication Using Edge-Coded Signaling

Shahzad Muzaffar • Ibrahim (Abe) M. Elfadel

Secure, Low-Power IoT Communication Using Edge-Coded Signaling

 Springer

Shahzad Muzaffar
Khalifa University
Abu Dhabi, United Arab Emirates

Ibrahim (Abe) M. Elfadel
Khalifa University
Abu Dhabi, United Arab Emirates

ISBN 978-3-030-95913-5   ISBN 978-3-030-95914-2   (eBook)
https://doi.org/10.1007/978-3-030-95914-2

This Springer imprint is published by the registered company Springer Nature Switzerland AG
The registered company address is: Gewerbestrasse 11, 6330 Cham, Switzerland

To my parents (Shamshad and Muzaffar),
beloved wife (Umber),
son (Anzar) and sisters (Shumaila, Shaista,
and Sumaira)
Shahzad

To the blessed memory of my brother
Zaher Elfadel
1965–2010
Abe

Prologue

*Your assumptions are your windows on the world.
Scrub them off every once in a while, or the light
won't come in.*

Isaac Asimov

The accepted IoT wisdom is that it is much more energy-efficient to compute on the edge than to communicate with the hub. Yet, this accepted wisdom is based on the premise that the power hog that lives in the communication subsystem of the IoT node must be avoided at all costs and that whatever power that is still available should be used to make the IoT node *smart*. The TinyML framework for machine learning on tiny micro-controllers is based on such premise and epitomizes the current wisdom under the edge-computing paradigm.

The ultimate aim of this research monograph is to revisit this premise by giving a concrete example of a novel, ultra-low-power, robust, and secure, IoT communication protocol that is meant to enable innovative IoT architectures that can bridge the chasm between edge and cloud computing.

The research described herein is a summary of several years of investigation into a single *what-if* question with regards to the design of signaling protocols, namely, *what if* the IoT communication subsystem can operate reliably and securely *without* the circuitry dedicated to clock and data recovery (CDR).

The main motivation of asking this *what-if* question is the basic observation that a CDR circuit is a significant contributor to power consumption in the communication transceiver. Being able to save as much of its power as possible in an IoT node is bound to impact the debate on computation vs. communication and on edge vs. cloud intelligence. In terms of real estate, transceivers with CDR circuits have tens of thousands of gates, and therefore a significant saving of silicon area will be achieved in case the CDR circuit is simplified or even possibly eliminated.

Not only do we give an existence proof of such a CDR-less communication link, but also, we provide a complete ecosystem of hardware and firmware built around such a communication link. This ecosystem comprises an application-specific processor, automatic protocol configuration, power and data rate management, cryptographic primitives, and automatic failure recovery modes. The resulting link and its associated ecosystem are fully compatible with IoT requirements on power, footprint, security, robustness, and reliability.

The fundamental idea of the proposed IoT communication protocol is to encode the ON bit in the data word as a sequence of pulses whose count is based on the ON-bit index. At the receiver, this index is decoded by simply counting the number of rising edges in the pulse sequence. This is the main reason we have called this protocol *Edge-Coded Signaling* or ECS.

From this basic idea, ECS has evolved through three different generations, ECS1, ECS2, and ECS3, into a full family of protocols. They are all variations on the fundamental theme of pulse generation for the ON bits at the transmitter end of the link and edge counting at the receiver end. They are all described in this monograph along with hardware prototypes that allow us to thoroughly benchmark and precisely quantify the IoT advantages of this novel family of signaling protocols. These advantages can be summarized as follows:

1. ECS results in a major simplification of the IoT device transceiver. This simplification in turn contributes to major gains in footprint, power savings, and cost.
2. ECS supports dynamic data rates, and the ECS parameters can be readily optimized to achieve the maximum average data rate for a given application.
3. ECS is robust in that it tolerates significant device-to-device variations in clock frequency as may be expected in a heterogeneous, asynchronous IoT network.
4. Along with a low-power design point due to transceiver simplification, ECS provides additional opportunities for power saving, both at the physical layer level and at the pulse design level, that are straightforward to implement.
5. ECS supports network protocols for automatic ECS parameter settings across a set of networked IoT devices. These automatic ECS configuration protocols are universal in that they can be applied for any network topology.
6. The ECS family of protocols is amenable to compact programming using a domain-specific, RISC-like, ECS processor. Its instruction set architecture achieves more than an order of magnitude of reduction in embedded code size and provides IoT designers with the flexibility to program new ECS protocols that are adapted to specific IoT communication tasks.
7. ECS supports low-overhead doubling of data rates using double-edge-coded signaling where both the rising and falling edge of the pulse are used to encode the ON bits in the bit stream.
8. ECS enables a close synergy between encoding and encryption, and provides a unique opportunity for significantly strengthening light-weight encryption algorithms in a way that is not possible with traditional signal encoding methods.

9. ECS development is supported with various tools for embedded C programming, debugging, and system integration. These tools greatly facilitate the deployment of hardware platforms for IoT sensor networks.
10. The robustness and reliability properties of ECS make it the signaling technique of choice in challenging media such as body-coupled communication.

The above 10 advantages are aligned with the 10 chapters of this monograph. By and large, each chapter is organized to lead from the IoT communication design problem to its solution under the ECS paradigm along with supporting hardware validation using either an FPGA or an embedded design platform. ASIC synthesis results using GLOBAFOUNDRIES 65 nm technology node have also been used throughout the chapters to further support the hardware results of the FPGA and embedded design platforms. Our own design of the ECS protocol targeted the sweet spot of a single-channel IoT communication link with a data rate in the range from 4.2 to 26.7 Mbps and with a power consumption cap of 20 μW.

Although significant work has already been invested in developing, testing, and validating ECS and its ecosystem, there are still several open research problems that are important to tackle in the next phase of ECS development. We have alluded to many of these problems at appropriate sections within the book chapters. In an epilogue to this book, we have consolidated and summarized all these open research problems with the hope that they will be of interest to colleagues and graduate students from the IoT research community.

Many of the results described in this monograph have already appeared in our prior conference and journal publications between 2015 and 2021. However, we have made a determined effort to synthesize these results and present them in a coherent notational and conceptual framework so that the monograph can serve as an accessible, self-contained reference, not just for IoT professionals but also for graduate students who are entering the field and interested in pursuing research in the area of secure, low-power IoT communication.

Abu Dhabi, United Arab Emirates   Shahzad Muzaffar
Abu Dhabi, United Arab Emirates   Ibrahim (Abe) M. Elfadel
November 2021

Acknowledgements

This research monograph is based on the PhD thesis of the first author conducted under the supervision of the second author at the Advanced Digital Systems Laboratory of the Masdar Institute, now part of Khalifa University, Abu Dhabi, UAE.

Several colleagues have contributed time, effort, and support to this research over the years. We particularly thank Dr. Jerald Yoo (National University of Singapore) and Dr. Ayman Shabra (MediaTek, USA) for helpful discussions at the early stages of this project. We also thank Dr. Zeyar Aung (Khalifa University, UAE) and Dr. Owais Waheed Talaat (Habib University, Pakistan) for their help with ECS encryption, and Mr. Numan Saeed (Mohamed Bin Zayed University of Artificial Intelligence, UAE) for his help with ECS automatic configuration. Special thanks are due to Prof. Neville Hogan (MIT) and Dr. Mihai Sanduleanu (Khalifa University, UAE) for serving on the PhD Thesis Committee of the first author and providing valuable feedback.

The authors gratefully acknowledge the support provided by the Semiconductor Research Corporation (SRC), USA, under the Abu Dhabi SRC Center of Excellence on Energy-Efficient Electronic Systems (ACE^{4}S), Contract 2013 HJ2440, with customized funding from the Mubadala Investment Company, Abu Dhabi, UAE.

They also thank the Office of Technology Management and Innovation at Khalifa University for their help in prosecuting US Patents 10,263,765 and 11,133,891.

Contents

Abbreviations

ACK	Acknowledgement
APD	Automatic Parameter Detector
ASIC	Application Specific Integrated Circuit
BAN	Boddy Area Network
BCC	Boddy Channel Communication
BER	Bit Error Rate
BPSK	Binary Phase Shift Keying
CDB	Clock Distribution Block
CDR	Clock and Data Recovery
Cflags	Combined Flags
CNOI	Combined NOIs
CoM	Center of Mass
CPM	Communication Processor Module
CPU	Central Processing Unit
DMA	Direct Memory Access
DS	Data Segment
Eb	Energy per Bit
ECS	Edge-Coded Signaling
ECSIA	Edge-Coded Signaling Interface Architecture
EDS	Encoded Data Segment
EFlags	Encrypted Flags
ENOI	Encrypted NOI
EPD	Encrypted PIC Data
ESC	Encoder and Select Control
ESC	Encoder and Selector
FPGA	Field Programmable Gate Array
FSK	Frequency Shift Keying
HBC	Human Body Communication
HDL	Hardware Description Language
IC	Integrated Circuit
ISA	Instruction Set Architecture

LSB	Least Significant Bit
LTC	Logical Topology Control
M2M	Machine-to-Machine
MA5/1	Modified A5/1
Mbps	Mega Bits Per Second
MCDCU	Multi-Core Debug Control Unit
MSB	Most Significant Bit
NOI/NOS	Number of Indices/Number of Symbols
NRZ	Non-Return-to-Zero
NST	Normal Serial Transfer
NVL	Number of Vulnerable Locations
OFDM	Orthogonal Frequency-Division Multiplexing
OOK	On-Off Keying
PCCU	PC Control Unit
PDG	Pulse and Delay Generator
PHY	Physical Layer
PIC	Pulsed-Index Communication
PIoT	Prototyped IoT
PLL	Phase-Locked Loop
PSR	Pulse Stream Receiver
RISC	Reduced Instruction Set Computer
RZ	Returns to Zero
SATA	Serial AT Attachment
SRL	Serializer
SSFC	Sandwiched Sensor Force Consolidators
ST	Schmitt Trigger
TCP/IP	Transmission Control Protocol/Internet Protocol
USB	Universal Serial Bus
VLSI	Very Large Scale Integration
WDM	Wavelength-Division Multiplexing

Chapter 1
Introduction

[The transmitter] could, for example, take a written message and use some code to encipher this message into, say, a sequence of numbers; these numbers then being sent over the channel as the signal.

Warren Weaver

Not only is the Internet of Things (IoT) extending the reach of the Internet of People (IoP) to the world of inanimate objects, but also it is providing innovators, engineers, and technologists, with a golden opportunity to revisit some of the fundamental assumptions that have been at the basis of the IoP physical infrastructure and evaluate their relevance and compatibility with the IoT physical infrastructure. One of these fundamental assumptions is that the receiver in a single-channel, serial communication interface will need a synchronization circuit in order to recover the data bits from the incoming bitstream. Examples of such single-channel, serial communication interfaces include the humble USB, the ubiquitous Ethernet, and the high-performance fiber optics network. The latter has of course been crucial for the exponential growth of the IoP. The receiver synchronization circuit goes with a name that perfectly describes its function: clock and data recovery (CDR). Indeed, each CDR has the double duty to infer the clock signal from the incoming data bits and use this inferred clock signal to retime (or resample) the data bits at the most appropriate instant of the bit time. The inference of the clock signal is based on the transitions experienced by the incoming data bits, while the sampling times are selected using feedback circuit architectures, the most common of which is the so-called phase-locked loop (PLL). A generic architecture of a CDR circuit using PLL is given in Fig. 1.1. For a thorough treatment of CDR architectures, the reader is referred to Chapter 9 of [67].

One important aspect of CDR performance is that it is very much dependent on the encoding of the incoming bits. One of the most common bit encodings is the Non-Return-to-Zero (NRZ) code in which bit 1 is coded HIGH and bit 0 is coded LOW throughout the bit time. One of the objectives of CDR is to generate a clock signal whose rate is equal to the bit rate. Such generation depends on the detection of transition *edges* between HIGH and LOW bits. When the incoming data has long

© Springer Nature Switzerland AG 2022
S. Muzaffar, I. M. Elfadel, *Secure, Low-Power IoT Communication Using Edge-Coded Signaling*, https://doi.org/10.1007/978-3-030-95914-2_1

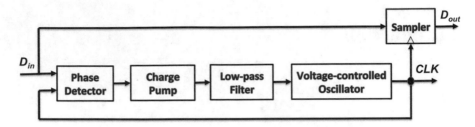

Fig. 1.1 Generic architecture of a clock and data recovery circuit using PLL

stretches of HIGH or LOW bits, the transition information is missing, and the PLL
has no edges to lock onto. One possible remedy to this loss of information is to
use bit encodings in which the 1 bit returns to zero (RZ) during the latter 50% of
the bit time, thus forcing a transition whenever a string of 1 bits is received. The
Manchester encoding forces transitions for both the 1 and 0 bits during bit time,
thus totally addressing the information loss problem in the CDR phase detector. Of
course, the disadvantage of such RZ encodings is that they require larger channel
bandwidth than the NRZ case or require that the transmitter send data bits at reduced
rates to meet channel bandwidth constraints. Another approach to balance the bit
transitions in the data stream is to use block encoding instead of bit encoding,
and one of the most widespread block techniques is the byte-oriented 8b/10b
encoding [79] that has been used in several communication technologies, including
the Gigabit Ethernet, USB 3.0, SATA, and InfiniBand, among many others. The
relevance of 8b/10b to CDR is that it guarantees that the difference between the
numbers of 1 bits and 0 bits in the 10-bit code is at most 2 and that no more than 5
bits can have the same value in a row.

Based on this well-established interplay between CDR performance and data
encoding, one is led to ask the extreme but basic question whether *there exist data
encodings that would totally do away with CDR while maintaining reasonable levels
of data rate and reliability*. The pertinence of this question to the IoT infrastructure
is that with such encodings the receiver architecture in a sensor or actuator device
can be significantly simplified with the resulting architecture taking much smaller
footprint and consuming less power. Of course, such simplified, low-power, compact
architecture must not compromise another crucial requirement for constrained IoT
nodes, namely, data authentication and security. One potential approach to providing
a layer of data protection is lightweight cryptography, which despite much recent
progress still requires a significant area and power overhead, while its lightweight
nature makes it a relatively easy target for malicious attacks.

The answer to this existence question has turned out to be affirmative with
the recent emergence of a novel family of block-oriented signaling techniques
for single-channel, serial communications. These techniques are based on the
fundamental concept of encoding data bits as pulse trains whose counts are also
transmitted and used by the receiver for decoding. Since the receiver uses the
rising edges of the pulses to decode the transmitted bits, this family of protocols

is called *Edge-Coded* Signaling (ECS) [56] and hence the title of this book. The three representative members of this novel family are ECS1 [57], ECS2 [51], and ECS3[56]. While the three techniques require no CDR, they differ in fundamental aspects that are related to transmission security and packet reliability. ECS1 and ECS3 have been proven compatible with symmetric stream ciphers and capable of providing multilayered protection of data in transmitted packets [59]. ECS2 has no such compatibility and is not amenable to multilayered encryption for secure data transmission. On the other hand, ECS2 and ECS3 achieve better data rates than ECS1 and have been experimentally proven to have less packet transmission failures than ECS1 [51]. We will explore the three ECS family member techniques in Chap. 2. The main objective of this chapter is to describe the most recent version of the family, ECS3, that combines and enhances the best advantages of ECS1 and ECS2 while avoiding their pitfalls and shortcomings. ECS1 and ECS2 are also included in the chapter as the earlier variants of the protocol. The newest family member shares with ECS1 and ECS2 the same underlying idea of encoding bits as pulse streams with inter-symbol spacings used to separate data words. However, ECS3 possesses several additional features above and beyond ECS1 and ECS2, including the following:

1. ECS3 employs an optimized segmentation process and a simplified encoding scheme that help reduce the number of ON bits and lower their index numbers. The process ultimately reduces the overall number of *packet* pulses needed to transmit a *data* word.
2. For a given data word length, ECS3 forms the most compact packet and therefore results in the maximum data rate. This is achieved mainly through a very compact header describing the encoding operations to which the data bits have been subjected.
3. Not only does ECS3 exploit the edge detection of received pulses to eliminate the need for CDR and to achieve remarkable robustness with respect to jitters, skews, and clock inaccuracies between the transmitter and receiver, but it also provides the flexibility of architecting transceivers to transmit multiple data words or to pipeline data transmission.
4. The ECS3 packet has a layered architecture and dynamic features that can be combined synergistically with the crypto algorithms to enhance communication security.
5. ECS3 is architecturally flexible in that it can be configured according to any signaling topology such as Master–Slave, Ring, Star, or Tree.

The data protection approach that ECS3 and ECS1 enable is based on a tight synergy between the communication protocol and the lightweight encryption algorithm with the communication protocol providing the encryption algorithm with additional protocol parameters whose encryption can further protect the data. In comparison with existing single-wire, CDR-less protocols such as 1-wire [10, 16, 26, 27, 44, 78], ECS1 has the double advantage of higher data rates (Mbps vs. Kbps) and stronger security (up to a factor of 2^{20} increase in attack complexity) while remaining within the IoT device envelop of power and footprint. The ECS security mechanism is discussed in detail in Chap. 7.

Chapter 3 focuses on the detailed analysis of ECS protocols. In this chapter, we also present a full quantitative analysis of the timing and robustness properties of ECS protocols, including the impact of important protocol parameters such as pulse width and inter-symbol separator on average data rate and protocol robustness with respect to clock variations. The main result of this chapter is a theoretical upper bound on clock variability between transmitter and receiver below which the protocol operates with zero decoding error over an ideal channel.

ECS encodes information using pulse counts with the counting based on one of the pulse edges. In Chap. 4, we address the problem of improving the ECS data rate for a given clock frequency and under a given power envelop by using both pulse edges of the ECS pulse stream. We call the novel protocol double data rate ECS (DDR-ECS) in analogy with DDR memory systems. While the concept is intuitive and attractive, its hardware implementation is not. This chapter, therefore, presents an efficient hardware design of the DDR-ECS transceiver that preserves the ECS built-in features while essentially doubling the data rate at the same clock frequency and within the same power budget [55].

Chapter 3 highlights the power savings that ECS can achieve as a result of the elimination of circuitry devoted to clock and data recovery. In Chap. 5, we show that further power saving can be achieved using the duty cycle of the pulse as a power control parameter. This power control policy is applied to a single-wire link with significant power saving achieved above and beyond the savings due to CDR elimination. These power savings are obtained without any impact on data rate [48].

Well-known multi-wire protocols, such as I2C, SPI, and UART, need to set the same baud rate on both ends of the link before the connected devices start communicating. The baud rates of all these devices are either factory set or require to be configured manually. Manual configuration is usually performed either by using the software settings or by hardware control. ECS family member techniques can also be configured on a per-device basis in a similar fashion, but such an approach will defeat the very purpose of ECS in providing a scalable, robust, ultra-low-power, and high data rate communication protocol for IoT devices. The error-free operation of ECS with maximum data rate requires a careful and judicious setting of ECS data packet and pulse timing parameters. In Chap. 6, we present a new algorithm for automatically detecting and setting the ECS protocol parameters at the power-on phase while removing the restriction on the IoT devices in the ECS network to communicate at a given baud rate. The hardware realization of the algorithm is power-efficient and uses closed-form formulas that assign suitable protocol parameters to both ends of the transmission link based on clock rate differences. This difference is determined by a preliminary exchange of clock pulse streams between the transmitter and the receiver. The automatic parameter setting remains operational even in the presence of variations between the local clock frequencies of the IoT devices communicating via ECS. The algorithm is illustrated in the case of several IoT devices with different local clock frequencies that are in need to synchronize their communication parameters with respect to the clock frequency of a master gateway node [58].

In Chap. 8, we present a domain-specific processor architecture, named Edge-Coded Signaling Interface Architecture (ECSIA). Aside from the traditional aspects of ISA design such as addressing modes, instruction types, instruction formats, registers, interrupts, and external I/O, the ISA includes domain-specific instructions that facilitate bit stream encoding and decoding based on the edge-coded signaling techniques. The domain-specific ECSIA micro-architecture employs a set of optimized processing blocks that can be used programmatically to encode and decode the transmitted data in the most economical way. The ECSIA allows customizations that support both standard edge-coded signaling techniques and specialized protocols that belong to the same family. The ECSIA design further allows an amalgamation of software and hardware that significantly reduces the number of instructions required to implement a given communication interface without impacting the data rates and reliability of the pulsed-signaling protocols [52, 53].

The last two chapters, Chaps. 9 and 10, are devoted to describing two recent applications of ECS communication technique. An FPGA hardware platform for the prototyping and analysis of ultra-low-power IoT sensor networks is discussed in Chap. 9. The platform is meant to address the problem of evaluating network topology design options for IoT sensor communications using single-channel communication protocols. The network topologies include bus, star, ring, and tree topologies. This FPGA-based IoT network platform is based on three fundamental ingredients: a full HDL implementation of the ultra-low-power TI MSP430 microcontroller, a novel ultra-low-power single-wire communication protocol that does not require any clock and data recovery, and embedded C implementation of the transceivers within the TI MSP430 without any need for external hardware circuitry. The platform is flexible in that it allows the design, analysis, and comparison of various networking graph topologies among the IoT sensors, including ones that contain gateways and hubs. The platform is also scalable in that the resources used for a two-sensor point-to-point communication are minimal [50].

The second application in Chap. 10 is a self-synchronizing, low-power, low-complexity body-coupled communication (BCC) transceiver using the ECS techniques. The unique features of these techniques are used to simplify the BCC transceiver hardware and reduce its power consumption by eliminating the need for circuits dedicated to clock and data recovery (CDR) and duty cycle correction. The self-synchronizing feature of the transceiver is achieved by exploiting the edge-coding property of ECS. A working prototype of the proposed BCC transceiver using off-the-shelf components is developed and used to test, for the first time, a full, bi-directional BCC link by transmitting arbitrary 16-bit data words through the human body over a range of $150\,cm$ with *zero* bit error rate and sub-$1nJ/bit$ energy efficiency [54].

Chapter 2
Edge-Coded Signaling Techniques

When one door closes another door opens; but we so often look so long and so regretfully upon the closed door, that we do not see the ones which open for us.

Alexander Graham Bell

The objective of this chapter is to present the analysis of signaling protocols for data transfer over a single-wire achieving high data rates (in the Mbs range), low power consumption, and small footprint. The protocols do not require a CDR, can operate with signals at low amplitude voltage ($\sim$1 V), has simple encoding and decoding schemes, and can tolerate baud rate differences between transmitter and receiver. We collectively refer to this new family as Edge-Coded Signaling (ECS) because its core idea is to transfer the indices of only the ON bits as a series of transition edges rather than bit times. A very compact packet header gives information about the number of such indices and the encoding operations to which the raw bits have been subjected. When the pulses are received, the receiver applies the appropriate decoding to infer the original data bits. The ECS protocols are dynamic in that they can accommodate several data rates. It exploits edge detection of incoming pulses to achieve remarkable robustness with respect to jitters, skews, and clock inaccuracies between the transmitter and the receiver. The protocols achieve significant improvements in data rate, reliability, packet security, and power efficiency with respect to state-of-the-art CDR-less techniques. ECS is also architecturally flexible in that it can be configured according to the signaling topology (Master–Slave, Ring, Star, etc.).

© Springer Nature Switzerland AG 2022
S. Muzaffar, I. M. Elfadel, *Secure, Low-Power IoT Communication Using Edge-Coded Signaling*, https://doi.org/10.1007/978-3-030-95914-2_2

2.1   Edge-Coded Signaling (ECS)

2.1.1   Edge-Coding Scheme

The core idea of ECS is to select the ON bits in a data word and transmit their
index numbers as pulse streams instead of transmitting the data bits themselves. An
example is given in Fig. 2.1 where the bit sequence "0101" (a) is transformed into
series of pulses (b) in which the count of pulses in each series is $n + 1$ with n being
the ordinal number of the ON bit in the binary sequence. In the example of Fig. 2.1b,
there are two series of pulses. The first series has one pulse corresponding to the
leading ON bit at position 0, and the second series has three pulses corresponding to
the ON bit at position 2. One series of pulses is separated from an adjacent one by an
inter-symbol separator α. Please note that α is not a time delay but rather a spacing
or separation symbol that is measured in clock cycles with the clock-cycle count
given by the local transmitter clock at transmission and the local receiver clock at
reception. The clocks at both ends do not have to be synchronized. Also, note that
one is always added to the pulse count corresponding to the index number. This

Fig. 2.1 (**a**) Standard serial
transfer. (**b**) Edge-coded
transmitter. (**c**) Edge-coded
receiver

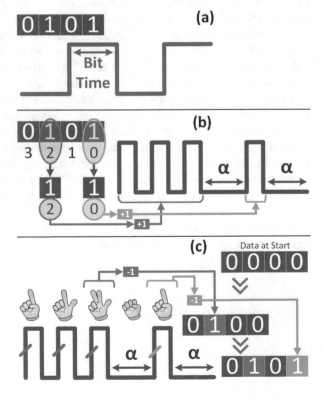

operation is necessary to handle the transmission of index 0. Otherwise, no pulse will be transferred if the bit at index 0 is ON. For each input pulse series, the ECS receiver counts the number of the incoming rising edges, subtracts one to retrieve the index number (i.e., $n = PulseCount - 1$), and sets a data bit at the index number. This is shown in Fig. 2.1c. The apparent drawback is that more work is seemingly needed to transmit such pulse series than the raw bits themselves. However, this is not the case as it is conceivable to achieve high data rates, using an encoding process that makes the index numbers as small as possible. This is accomplished by breaking the bit stream into smaller segments, reducing the number of ON bits as much as possible in each segment and relocating these ON bits to the lowest index positions. The encoding information and the number of ON bits in the encoded data are sent as a packet header along with the index numbers. All the information in the packet header itself is transmitted as pulse streams, exactly as the index numbers. In short, instead of transmitting bits, ECS codes them as edge counts and transmits them along with the formatting information, itself edge-coded, so that the receiver is able to reconstitute the data word. The steps involved in ECS transmission are explained in the following subsections.

2.1.2 ECS Segmentation

The number of pulses to transmit increases rapidly with the data word size B and the number of its ON bits. The most significant bits require larger number of pulses to represent their index numbers. Considering the worst case where all the bits are ON, the number of pulses required would be $B(B + 1)/2$. The rapid increase in the number of pulses reduces the data rate rapidly and, therefore, the count of pulses must be limited. To do so, ECS breaks the data word into smaller segments of size $l = 4$ bits each, thus limiting the index numbers to a maximum of 3 (i.e., $4 + 3 + 2 + 1$). With the reduced segment size, the maximum number of pulses per segment reduces to 10. An ECS segmentation example is given in Fig. 2.2 where a 16-bit word is partitioned into 4-bit segments $S_i, 1 < i < 4$. The segmentation steps are also shown in Algorithm 1 on lines 1 and 2. The optimization of the segment size l is discussed in Sect. 2.2.

2.1.3 ECS Encoding

The increase in data word size also increases the number of inter-symbol separators needed to separate out the pulse streams, representing the ON bits. Such separators reduce the data rate significantly. Reducing the number of ON bits helps in mitigating the effect of separators on data rate. ECS encoding effectively reduces the number of ON bits in each data segment. The ECS encoding is simply a conditional bit-wise NOT operation on a target segment, with the condition being that the

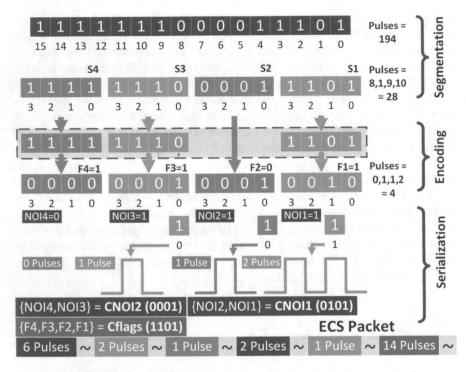

Fig. 2.2 Example: ECS packet formation

Algorithm 1 ECS segmentation and encoding process

Inputs:
- Data: 16-bit data word
Outputs:
- $CFlags$: Concatenated Encoding Indicators
- $CNOI_1, CNOI_2$: Concatenated ON-bit Counts

1: $S_1 = Data[3:0]$, $S_2 = Data[7:4]$
2: $S_3 = Data[11:8]$, $S_4 = Data[15:12]$
3: **for** i=1 **to** 4 **do**
4:      $NOI_i = countONbits(S_i)$
5:      $F_i = 0$
6:      **if** $NOI_i > l/2$ **then**
7:          $S_i =\sim S_i$
8:          $F_i = 1$
9:          $NOI_i = countONbits(S_i)$
10:     **end if**
11: **end for**
12: $CFlags = \{F_4, F_3, F_2, F_1\}$
13: $CNOI_1 = \{NOI_2, NOI_1\}, CNOI_2 = \{NOI_4, NOI_3\}$

---

number of ON bits in a segment is longer than half of the segment size. To explain the encoding scheme further, let us assume $B = 16$. If a segment satisfies the said condition, bits of the segment are inverted and a 1-bit flag, F_i, is set to represent the applied operation. The subscript i represents the segment number. The encoding steps are presented in Algorithm 1 from lines 3–11 and in the encoding section of Fig. 2.2. Each segment is processed independently, and four distinct flags are generated, one for each segment. These four flags are then concatenated to yield a single 4-bit flag named $CFlags$. Additionally, the encoding process produces four 2-bit *Numbers of Indices*, each, denoted NOI_i, representing the number of ON bits in segment i. The $NOIs$ of two consecutive segments are concatenated to yield two 4-bit *concatenated NOIs* that are denoted $CNOIs$. The generation of $CFlags$ and $CNOIs$ is shown in Algorithm 1 on lines 12 and 13 and in the serialization section of Fig. 2.2. At the end of this process, all the information required for transmission gets compacted in nibbles of 4 bits each, which is the same as the size of the 4-bit segment and, hence, helps in maximizing the data rate. The efficient hardware implementation of an encoder performing segmentation and encoding is discussed in Sect. 2.4.

2.1.4 ECS Transmitter

Pulse Stream and Separator Generation Scheme

In the ECS transmission process, the ECS scheme for generating the pulse streams and the inter-symbol separators plays a crucial role. The encoding pulses in the ECS packet and the α spacings between packets are generated using the ECS clock, which can be obtained in two ways. One way is for the system clock to be routed directly to the ECS clock port. Another way is for the system clock to be divided to generate a slower ECS clock. The pulse generation process is illustrated in Fig. 2.3. The ECS clock is ANDed with a control signal, *Pulse Stream Active* (PSA), set by the control module. The PSA is high during the transmission of a pulse stream, allowing the ECS clock cycles to go through. During the transmission of the inter-symbol separator α, PSA is low, thus gating the ECS clock. Please note that α in ECS is not a time delay, but rather a count of the rising or falling edges of the ESC clock. In Fig. 2.3, we have used $\alpha = 4$ clock cycles as it is the optimal count at which the maximum data rate is achieved. This will be discussed further in Sect. 2.2.

Transmission Flow

The format of the ECS packet is shown in Fig. 2.4 and a numerical example is given in Fig. 2.2. The $CFlags$ are transmitted to inform the receiver about the encoding process, while the $CNOIs$ are transmitted to help the receiver account for all the incoming ON bit indices. The ECS transceiver starts the transmission by sending a

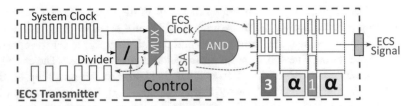

Fig. 2.3 ECS pulse stream and inter-symbol separator generator (indices from Fig. 2.1)

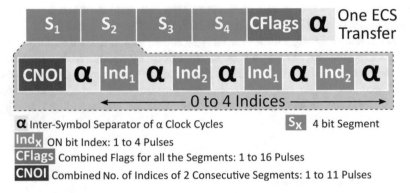

Fig. 2.4 ECS packet

pulse stream with a pulse count equal to $CNOI_1 + 1$ followed by an inter-symbol separator α of four clock cycles (line 2 in Algorithm 2). The additional pulse is needed to inform the receiver not to expect any index number when the count of ON bits in the first two segments is zero. Next, the transmitter sends a number of pulse streams equal in count to $NOI_1 + NOI_2$ followed by an α at the end of each stream. The pulse count in each pulse stream is equal to the index number of an ON bit in segments S_1 and S_2 plus an additional pulse, making a total of $index + 1$ pulses. The additional pulse is used to handle the transmission of a zero index number. The transmission process of indices is presented in Algorithm 2 on lines 3 through 9. A similar transmission follows for the next two segments, S_3 and S_4, during which $CNOI_2$ and the index numbers of the ON bits in these segments are transmitted. At the end of the transmission of all segments, the CFlags are transmitted, also in the form of a pulse stream followed by an α (line 10 in Algorithm 2). Their pulse count is equal to $CFlags + 1$. An additional pulse is needed to represent zero content of $CFlags$ as in the case when no segment goes through the encoding inversion. The graphical transmission process and the generated waveforms are shown in Fig. 2.6.

Algorithm 2 ECS transmitter algorithm

Inputs:
- $CFlags$: Concatenated Encoding Indicators
- $CNOI_1$, $CNOI_2$: Concatenated ON-bit Counts **Outputs:**
- ECS Signal: The pulse streams and inter-symbol separators

1: **for** j=1,2 **do**
2:     sendPulsesWithSeparator($CNOI_j + 1, \alpha$)
3:     **for** each ON bit in S_{2j-1} with index i **do**
4:         sendPulsesWithSeparator($i + 1, \alpha$)
5:     **end for**
6:     **for** each ON bit in S_{2j} with index i **do**
7:         sendPulsesWithSeparator($i + 1, \alpha$)
8:     **end for**
9: **end for**
10: sendPulsesWithSeparator($CFlags + 1, \alpha$)

2.1.5 ECS Receiver

Pulse Stream and Separator Reception

The ECS receiver is unique in that it does not require any clock and data recovery (CDR) circuitry either to receive the incoming data over single channel or to synchronize it with a local clock. The ECS exploits detection and count of edges of the incoming pulse streams to receive all the information required to rebuild the transmitted data successfully. Contrary to standard serial transfer, the width of transmitted pulses is inconsequential to ECS and, hence, does not employ this information in receiving data. Though there could be different implementations of the ECS receiver, our implementation is focused on counting the number of clock cycles between two ECS pulses to detect the inter-symbol spacing and separate out the incoming pulse streams. The ECS pulse stream reception process is illustrated in Fig. 2.5. The ECS receiver keeps track of two counts, the pulse count and the clock count. The reception process starts with the very first rising edge of the input pulse stream. At each rising edge of the pulse stream, the pulse count is incremented, and

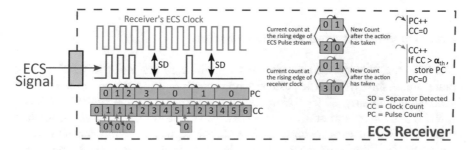

Fig. 2.5 ECS pulse stream and inter-symbol separator receiver (input signal from Fig. 2.3)

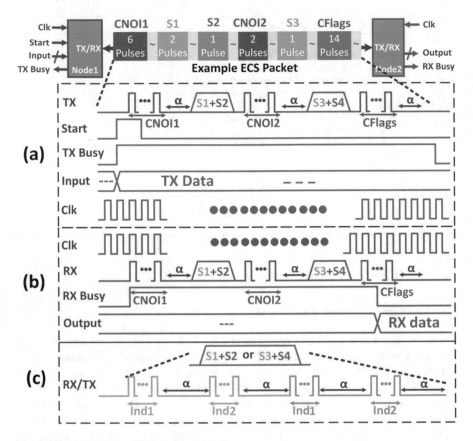

Fig. 2.6 (**a**) Transmitter. (**b**) Receiver. (**c**) Indices

the clock count is cleared. At each rising edge of the receiver's ECS clock, the clock count is incremented and compared with a separator threshold α_{th} that is set to half of α in our implementation of the protocol. If the condition is satisfied, an inter-symbol separator is declared, and the current pulse count is accordingly stored as a record of the transmitted packet.

Reception Flow, Decoding, and Reconstruction

The ECS packet reception process starts with the very first rising edge of the first pulse stream for $CNOI_1$. The pulse stream is received as described in the previous subsection. Similarly, all the following pulse streams for the indices of S_1, S_2, and $CNOI_2$ and the indices of S_3, S_4, and $CFlags$ are received, and the corresponding parts of the ECS packet are updated. Bits of each segment are complemented if the corresponding bit in $CFlags$ is set. At this stage, all the transmitted segments are assembled to rebuild the data word. The full process of receiving, decoding, and assembling is shown in Algorithm 3. An efficient implementation of ECS encoding and decoding is discussed in Sect. 2.4. The transmission and reception processes along with their generated waveforms are shown in Fig. 2.6.

Algorithm 3 ECS receiver algorithm

Inputs:
- ECS Signal: The pulse streams and inter-symbol separators
Outputs:
- Data: 16-bit data word

1: **for** i=1 **to** 2 **do**
2:     $CNOI_i$ =PulseStreamReceiver()*-1
3:     $NOI = CNOI_i[1:0]$
4:     $S_{2i-1} = S_{2i} = 0$
5:     **for** j=1 **to** NOI **do**
6:         $index$ =PulseStreamReceiver()-1
7:         $S_{2i-1}[index] = 1$
8:     **end for**
9:     $NOI = CNOI_i[3:2]$
10:    **for** j=1 **to** NOI **do**
11:        $index$ =PulseStreamReceiver()-1
12:        $S_{2i}[index] = 1$
13:    **end for**
14: **end for**
15: $CFlags$ =PulseStreamReceiver()-1
16: $Data = \{S_4 \oplus \{4\{CFlags[3]\}\}, S_3 \oplus \{4\{CFlags[2]\}\}, S_2 \oplus \{4\{CFlags[1]\}\}, S_1 \oplus \{4\{CFlags[0]\}\}\}$
17:

---

* PulseStreamReceiver() is the pulse counter for each input pulse stream (Fig. 2.5)

2.1.6 ECS Transmission System

The ECS communication technique can be used with a variety of channels such as wired, wireless, infrared, and human body channel. ECS is advantageous to all these channels as it results in significant simplification of transceiver circuitry, reduction in power consumption, and decrease in footprint. A simple PHY layer for single-wire communication is shown in Fig. 2.7a where two tri-state buffers are used to switch channel access between transmitter and receiver. Moreover, ECS can be used with any communication medium without any significant change, as shown in Fig. 2.7b. In case of wireless transmission, the wireless front end can be easily used to transmit and receive the packet pulses. As ECS does not need power-hungry circuits such as CDR or duty cycle correction, the complexity of the front end reduces significantly as compared to standard transceivers. Additionally, ECS helps in improving the bit rate of wireless transmission. For example, the standard OOK and ASK modulation techniques need duty cycle accuracy to recover square pulses. As the transmission data rate increases, the output pulses turn into triangular

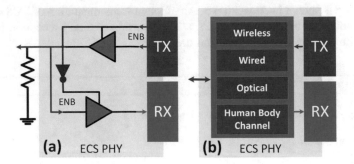

Fig. 2.7 (**a**) ECS PHY for single-wire. (**b**) General ECS PHY block diagram

ones at the receiver end of the wireless modules, which limits the bit rate. On the other hand, ECS does not depend on the duty cycle and can correctly decode the triangular and demodulated pulses as long as their peaks remain above the detection threshold of the ECS receiver. This is also the case with infrared channels where the power consumption and complexity of the optoelectronic system are reduced. ECS can also help in reducing the transceiver complexity of human body-channel communication (BCC) [69] by eliminating duty cycle dependency and the need for CDR while enabling the processing of pulses deformed through the variable-gain human body channel. ECS has been successfully tested with all these channels, and the experiments are discussed in Sect. 2.4.

2.1.7   ECS Data Rate

Let b_i^s be the i-th bit in the s-th encoded segment and l be the number of bits per segment. The total number of segments, N, is given by

$$N = \frac{B}{l} \tag{2.1}$$

In the ECS packet, the $CNOIs$, $CFlags$, and segments all have the same length l. Therefore, each $CFlag$ represents l consecutive segments among a total of N segments. The number of $CFlags$, n_{cf}, in the ECS packet is given by

$$n_{cf} = \frac{N}{l} \tag{2.2}$$

Similarly, each $CNOI$ concatenates the $NOIs$ of 2 consecutive segments among a total of N segments. The number of $CNOIs$, n_{cn}, in the ECS packet is therefore given by

$$n_{cn} = \frac{N}{2} \tag{2.3}$$

For a segment s, the required number of pulses is given by

$$P_s = \sum_{i=0}^{l-1}(i+1)b_i^s \tag{2.4}$$

and the number of ON bit indices is given by

$$NOI_s = \sum_{i=0}^{l-1} b_i^s \tag{2.5}$$

Let PI_x be the number of pulses required for one $CNOI$. The PI_x is given as

$$PI_x = 1 + NOI_{2x-1} + 2^{l/2}NOI_{2x} \;,\quad 1 \le x \le n_{cn} \tag{2.6}$$

where one additional pulse is used to represent the absence of ON bits. The x subscript in (2.6) refers to two consecutive segment numbers, one odd and one even, for NOI_s in (2.5). Now, let PF_y be the number of pulses required for one $CFlags$. PF_y is given as

$$PF_y = 1 + \sum_{i=0}^{l-1} 2^i F_s \;,\quad s = i + l(y-1) \;,\quad 1 \le y \le n_{cf} \tag{2.7}$$

where F_s is the flag bit for the s-th encoded segment. Again, one additional pulse is used to represent the *no-encoding* state.

The number of pulses for $CNOIs$, $Segments$, and $CFlags$ and the total number of NOI pulses are, respectively, given by

$$n_{pi} = \sum_{x=1}^{n_{cn}} PI_x \tag{2.8}$$

$$n_{ps} = \sum_{s=1}^{N} P_s \tag{2.9}$$

$$n_{pf} = \sum_{y=1}^{n_{cf}} PF_y \tag{2.10}$$

$$n_{in} = \sum_{s=1}^{N} NOI_s \tag{2.11}$$

The total pulse count is therefore given by

$$C = \left(n_{cf} + n_{cn} + n_{in}\right)\alpha + n_{pi} + n_{ps} + n_{pf} \tag{2.12}$$

where α is the number of clock cycles for an inter-symbol separator. The data rate R of the ECS protocol for a bit stream of B bits, clock period of T, and a total pulse count of C is given as

$$R = \frac{B}{TC} \tag{2.13}$$

The optimum values of the protocol parameters are derived in Sect. 2.2.

2.2   ECS Optimizations

2.2.1   Optimum Inter-symbol Separator α

Since all the transmitted pulse streams are separated by an inter-symbol separator, an appropriate value of α is indispensable for successful packet reception and for maximizing data rate. Keeping all the parameters in (2.12) and (2.13) constant except for α, we obtain the relationship $R \propto a/(b + c\alpha)$, where a, b, and c are constants. This relationship clearly shows that an increase in α decreases data rate, as shown in Fig. 2.8a. Both empirically and theoretically [49], the smallest value of α for guaranteeing correct decoding is 4 clock cycles [49]. Below this value, the receiver would fail to decode the packet successfully because of the ambiguity between pulse spacing and inter-symbol separators. A value of α larger than 4 will increase the tolerance to clock variations but decrease data rate and reduce reliability with respect to packet failure.

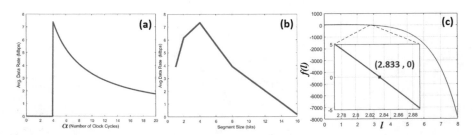

Fig. 2.8 (**a**) Data rate vs. α. (**b**) Data rate vs. segment size. (**c**) $f(l)$ vs. segment size l (Eq. (2.18))

2.2.2 Optimum Segment Length l

To find the optimum segment length that maximizes data rate, we minimize the number of clock cycles needed to transmit the ECS packet. We know from the previous section that the number of segments is $N = B/l$. Assuming that bits 0 and 1 are equally likely, the expected value of P_s in (2.4) is

$$E[P_s] = \frac{l(l+1)}{4} \tag{2.14}$$

Similarly, the expected value of PI_x in (2.6) is

$$E[PI_x] = \frac{2 + l(1 + 2^{l/2})}{2} \tag{2.15}$$

Assuming the flag bit F_s is equally likely to be 0 or 1, the expected value of PF_y in (2.7) is

$$E[PF_y] = \frac{1 + 2^l}{2} \tag{2.16}$$

Using $N = B/l$, the expected value of the total number of clock cycles C as given in (2.12) becomes

$$E[C] = \left(\frac{B}{l^2} + \frac{B}{2l} + \frac{B}{2}\right)\alpha + \frac{B}{2l} + \frac{B(1 + 2^{l/2})}{2} + \frac{B(l+1)}{4} + \frac{B(1 + 2^l)}{2l^2} \tag{2.17}$$

Taking the derivative with respect to l and equating it with zero (i.e., $\partial E[C]/\partial l = 0$), we get

$$f(l) \triangleq \frac{\partial E[C]}{\partial l} = \alpha(8 + 2l) + 2l - l^3(2^{l/2}\ln(2) + 1)$$

$$-2^l(2l\ln(2) - 4) + 4 \tag{2.18}$$

$$= 0$$

A graphical method to find the optimal segment length l_{opt} for a given α is to plot $f(l)$ as function of l and find the l intercept point. Such plot is shown in Fig. 2.8c for $\alpha = 4$, which results in

$$l_{opt} = 2.833 \approx 3 \quad bits \tag{2.19}$$

ECS divides the data word into segments of equal size and, therefore for a word size that is a power of 2, there are two possibilities of optimum segment length, 2 and 4. We select $l_{opt} = 4$ because there is a negligible reduction in data rate as compared to $l = 2$ where the degradation is significant. If the segment length is increased or decreased from the optimum value of 4, the data rate degrades rapidly, as shown in Fig. 2.8b. Segments smaller than 4 bits reduce the data rate due to the increased number of inter-symbol separators. In contrast, segment lengths larger than 4 bits affect the data rate negatively due to the increase in most significant bit (MSB) index numbers. To achieve maximum data rate, the protocol must be operated with segments of length 4 bits each.

2.3   Earlier Versions of ECS

ECS1 and ECS2 are earlier versions of the ECS3 protocol described in the previous sections. With slight differences, these techniques apply an encoding scheme to a data word B to *minimize* the number of ON bits and *move* them to the least significant bit (LSB) end of the packet with the goal of lowering the number of pulses required to transmit the data bits. The encoding process includes a segmentation step where the data is broken into N independent segments of size l bits each (i.e., $N = B/l$). To maximize data rate, they use, on each segment, an encoding combination of bit inversion and/or segment reversion/flipping. For ECS1, this combination is meant to reduce the number of ON bits and decrease their index values. For ECS2, the same combination is intended to reduce the number of ON bits and decrease the decimal number represented by each segment. To facilitate decoding, flag pulses representing the type of encoding performed are added to each segment. Unlike ECS1, the ECS2 segment flags of two consecutive segments are combined in one data word flag and placed in the header. ECS2 further applies a third segmentation step post-encoding, the level-2 segmentation, whose goal is to further reduce the number of pulses per segment and, therefore, increase the data rate.

All the pieces of information including flags, the number of indices, and the indices themselves in the case of ECS1, or the decimal numbers of each segment in the case of ECS2, are transmitted in the form of pulse streams. The pulse is characterized by its width which is the number of clock cycles during which it remains high. Within a given packet, segment pulse streams are separated by an inter-symbol separator α. The ECS1 and ECS2 packet formats are presented in Figs. 2.9 and 2.10. To describe the process of ECS1 and ECS2 data transmission, examples are given in Figs. 2.11 and 2.12, respectively. A decimal number 65,055 is considered as a 16-bit data word for transmission. The 16-bit data word is divided into two independent segments, each of 8 bits, which reduces the index numbers of MSB bits and, consequently, the number of pulses to represent the ON bits. Because the number of ON bits in Segment#1 is higher than half of the segment length (5 and 4, respectively), the bits are inverted, and the Flags of Segment#1 are set to 2. This step further reduces the number of ON bits in Segment#1, but the index numbers of the ON bits are located in the MSB part of the segment. The bit-wise flip operation is therefore applied to relocate the ON bits to the LSB part, which results in the

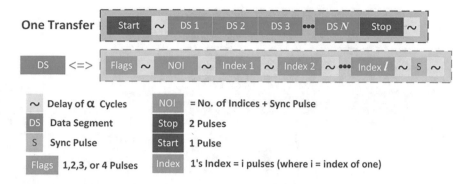

Fig. 2.9 ECS1 packet format

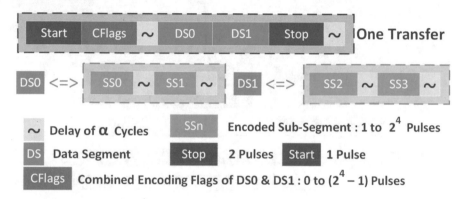

Fig. 2.10 ECS2 packet format

reduction of the number of pulses. The *Flags* field of Segment#1 is now updated to 3, signifying that both of the encoding operations are applied. The same steps are applied to Segment#2 except that only the inversion operation is needed. The *Flags* field of Segment#2 is set to 2, signifying that only the inversion operation is applied. In the case of ECS1, all the packet information including the encoded segments, *Flags*, and the number of ON bit locations (*NOIs*) is now available to start the transmission. However, in the case of ECS2, an additional segmentation step is applied where each encoded segment is divided into two sub-segments. All the pieces of information are transmitted in the form of pulse streams separated by inter-symbol separators.

The receiver counts the number of pulses for each pulse stream and applies the decoding according to the *Flags* field in the received packet. Like ECS3, ECS1 has a variable number of symbols per data word, which enables the addition of security layers, whereas ECS2 presents a fixed number of symbols per data word, which improves transmission reliability with respect to packet failures. The three protocols of the ECS family are compared in Table 2.1 for a 16-bit data word transmission.

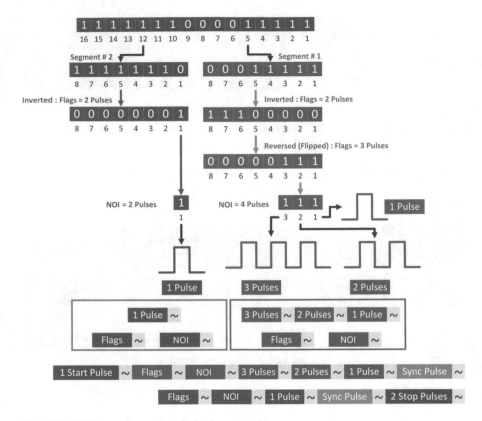

Fig. 2.11 ECS1 encoding and packetization example

2.3.1   Data Rates

ECS1 and ECS2 share with ECS3 the same notation and data rate equations as given in Sect. 2.1.7. However, the mathematical definitions of some of the symbols may vary as per the differences in their packet structure. All these variations are presented in Table 2.2. The generalized data rate equations for the ECS family are shown in Rows 14 and 15. Therein, the symbol n_{pe} represents configuration pulses that include start, stop, and sync pulses.

2.3.2   Optimizations

The segment size is chosen to maximize data rate. For a small segment, the inter-symbol separators inserted between pulse streams to separate symbols reduce the data rate. Similarly for large segments, ON bits with high indices require a large

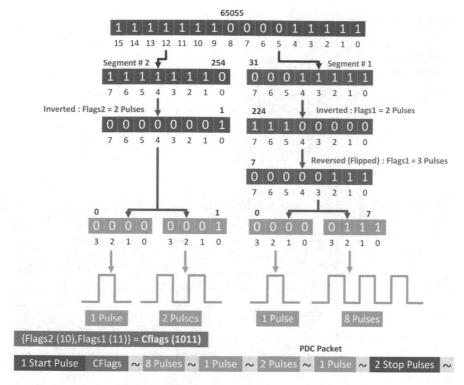

Fig. 2.12 ECS2 encoding and packetization example

number of pulses to be transmitted, which in turn reduces the data rate. It is therefore intuitive that there is a segment size for which the data rate is maximum. For ECS1, the data rate is maximized when the number of bits per segment is 8. For ECS2, the level-2 sub-segment (SS) size is appropriately chosen to maximize the data rate. The number of pulses increases exponentially with the size of SS, which in turn drastically reduces the data rate. For ECS2, the data rate is maximized when the number of bits per sub-segment is 4. The process of finding the optimum segment size is the same as the one presented in Sect. 2.2.2.

2.4 Experimental Setups and Results

An ECS communication system is implemented in Verilog HDL over Xilinx Spartan-6 FPGA board and verified through simulation and real-time communication between two nodes. For an apple-to-apple comparison with the earlier versions of the protocol, similar communication systems are developed for ECS1 and ECS2. However, this section describes only the ECS implementation. The development of the ECS system includes the efficient hardware implementation of ECS encoder

Table 2.1 Comparison of ECS family member techniques using 16-bit data word

		ECS1	ECS2	ECS3
Packetization				
Segment size	Level 1	8	8	4
	Level 2	–	4	–
No. of segments	Level 1	2	2	4
	Level 2	–	4	–
CFlags	Count	2	1	1
	Size (bits)	2	4	4
CNOIs	Count	2	–	2
	Size (bits)	2	–	4
Encoding steps		Invert + Flip	Invert + Flip	Invert
No. of symbols/packet		Dynamic	Fixed	Dynamic
Performance (25 MHz clock, 65nm CMOS technology)				
Data Rate	(Mb/s)	3.1–8.5 (4.1)[a]	4.8–12.9 (7.3)	4.2–26.7 (6.4)
	(% ↑)[d]		54.8–51.7 (78)[a]	35.5–214 (56)
Power	(µW)	≈26.6	≈25	≈19
	(% ↓)[d]		6	28.5
Eb	(pJ/bit)	3.1–8.5 (6.5)[a]	1.9–5.1 (3.4)	0.7–4.5 (2.9)
	(% ↓)[d]		38.7–40 (47.7)[a]	47–77.4 (55.4)
Area	(*Gatecount*)	≈2356	≈2150	≈2098
	(% ↓)[d]		8.7	10.9
Security[b]		Yes	No	Yes
Reliability	(NVLs)[c]	10–18	6	5–12
	(% ↑)[d]		up to 66.7	up to 72.2

[a] (Avg.)
[b] Packet protection
[c] No. of vulnerable locations
[d] Improvement relative to ECS1

and decoder. Both the encoder and decoder are combinatorial in nature and present a low power operation without any extra computational overhead. The encoder is implemented as a single hardware block that works with one 4-bit segment as input and generates the corresponding NOI, F_s, and the encoded segment. The encoder truth table is shown in Table 2.3, where S is the input data segment and S_E is the encoded segment at the output. Due to the segmentation and encoding process, there is a maximum of two ON bits per segment for which the index numbers Ind_1 and Ind_2 need to be transmitted. The ECS decoder at the receiver end takes as input all the received index numbers for a given segment and outputs a 4-bit data segment.

Table 2.2 Data rates of ECS member techniques

	Parameter	Notation	ECS1	ECS2	ECS3
1	Segment size		l	$l_1, l_2, l = l_2$	l
2	No. of segments	N	B/l	$N_{l_1} = B/l_1$ $N = N_{l_2} = \frac{N_{l_1}}{2} = \frac{B}{l_1 l_2}$	B/l
3	No. of CFlags	n_{cf}	N	$N_{l_1}/2$	N/l
4	No. of CNOIs	n_{cm}	N	0	$N/2$
5	Pulses/segment	P_s	$\sum_{i=0}^{l-1}(i+1)b_i^s$		$\sum_{i=0}^{l-1} b_i^s$
6	On bits/segment	NOI_s	$\sum_{i=0}^{l-1} b_i^s$	0	$\sum_{i=0}^{l-1} b_i^s$
7	Pulses/CNOI	PI_x^a	$1 + NOI_x$	0	$1 + NOI_{2x-1} + 2^{l/2} NOI_{2x}$
8	Pulses/CFlags	PF_y^b	$1 + f_0^y + 2f_1^y$	$1 + f_0^y + 2f_1^y + 4f_2^y + 8f_3^y$	$1 + \sum_{i=0}^{l-1} 2^i F_s$ $s = i + l(y-1)$
9	Total CNOIs pulses	n_{pi}	$\sum_{x=1}^{n_{cn}} PI_x$		
10	Total segment pulses	n_{ps}	$\sum_{s=1}^{N} P_s$		
11	Total CFlags pulses	n_{pf}	$\sum_{y=1}^{n_{cf}} PF_y$		
12	Total ON bits	n_{in}	$\sum_{s=1}^{N} NOI_s$	N	$\sum_{s=1}^{N} NOI_s$
13	Total extra pulses	n_{pe}^c	$3 + 2\alpha + N(2 + \alpha)$	$3 + 2\alpha$	0
14	Total pulse count	C	$(n_{cf} + n_{cn} + n_{in})\alpha + n_{pi} + n_{ps} + n_{pf} + n_{pe}$		
15	Data rate	R	B/TC		

[a] $1 \le x \le n_{cn}$
[b] $1 \le y \le n_{cf}$
[c] Start/stop and sync pulses

Table 2.3 ECS encoder

S	S_E	F	NOI	Ind_2	Ind_1
0000/1111	0000	0/1	00	000	000
0001/1110	0001	0/1	01	000	001
0010/1101	0010	0/1	01	000	010
0011	S	0	10	010	001
0100/1011	0100	0/1	01	000	011
0101/0110	S	0	10	011	001/010
0111/1000	1000	1/0	01	000	100
1001/1010/1100	S	0	10	100	001/010/011

The equations of the ECS decoder logic are

$$\{C, B, A\} = Ind_1 \tag{2.20}$$

$$\{F, E, D\} = Ind_2 \tag{2.21}$$

$$DS_1 = \{\bar{C}, B \cdot A, B \cdot \bar{A}, \bar{B} \cdot A\} \tag{2.22}$$

$$DS_2 = \{\bar{F}, E \cdot D, E \cdot \bar{D}, \bar{E} \cdot D\} \tag{2.23}$$

$$S_s = (DS_2 | DS_1) \oplus \{F_s, F_s, F_s, F_s\} \tag{2.24}$$

where DS_0 and DS_1 denote intermediate Verilog wires.

The ECS experimental setup comprises two nodes, each of which employs the abovementioned encoder and decoder along with the transmitter and receiver algorithms, respectively, all implemented in Verilog HDL. The ECS Transmission Algorithm 2 and Reception Algorithm 3 are synthesized as finite state machines. We have used 16-bit data words at a clock rate of 25 MHz. The transmitter at the first node sends a 16-bit data starting at 0 with an increment of 1 at each transmission. The second node resends the same data back. The returned and original data words are compared to verify the complete round-trip chain, and the number of perfect matches is logged. The ECS technique is verified using a number of single-channel links such as single-wire, wireless, infrared, and human body channel. In wired communication, a single-wire is used to connect both nodes using the PHY layer shown in Fig. 2.7. For wireless communication, a 433 MHz transceiver is used which accepts a raw ECS bit stream and transmits it wirelessly using OOK/ASK modulation. Similarly, for infrared communication, a simple infrared transceiver driver circuitry is used. For human body channel communication, new transceivers have been developed to carry out transmission through the human body. In all four experiments, ECS achieves flawless transmission. It must be noted that to support the aforementioned communication channels, ECS remains unchanged and only the front ends are replaced to transfer pulse streams through the desired channel. As highlighted in Sect. 2.1, ECS improves the bit rate of wireless transmission, which is verified with the experimental setup under discussion. Indeed we have observed an increase in bit rate from 4.8 to 20 Kbps ($\approx$300% improvement). Similar observations have been recorded while testing the infrared communication channel. For body channel communication, ECS helps in reducing the transceiver complexity, which is a significant advantage. In this latter case, a cascade of an amplifier, a discrete filter, and a level detector is sufficient to recover the ECS pulses traveling through the human body [54].

Along with the FPGA prototype, we have also synthesized and verified the ECS3 design using a Synopsis logic-synthesis flow and a GLOBALFOUNDRIES 65 nm process in order to get the most realistic area and power estimates and compare them with the published literature. We have determined that ECS3 consumes 19 μW with a gate count of 2098 gates, offering dynamic data rates in the range of 4.2–26.7 Mb/s (averaging 6.4 Mb/s) with a 25-MHz clock. As mentioned above, the selection of the 25 MHz clock rate is just for illustration purposes. We have verified the functionality of ECS3 using frequencies in the range from few KHz up to 200 MHz, the maximum frequency supported by our FPGA platform. As is clear from (2.13), the ECS3 data rate increases linearly with the clock frequency, and higher clock frequencies can be used to achieve higher data rates. Compared with NRZ serial transfer (NST) using CDR, ECS3 reduces area and power by more than 87% and 78%, respectively. Table 2.1 summarizes and compares the performance parameters of ECS1, ECS2, and ECS3. The data rate of ECS3 is increased significantly as compared to ECS1 and is as good as ECS2. ECS3 consumes less power and is more energy-efficient than ECS1 and ECS2 while maintaining a small footprint. Additionally, ECS3 helps in providing packet security, as will be discussed in Chap. 7. The reliability of ECS3 is similar to ECS2. It provides an improvement of up to 72.2% as compared to ECS1. In a nutshell, the results show that, overall, ECS3 outperforms both ECS1 and ECS2. Table 2.4 compares ECS3 with NST, which includes CDR, in terms of area and power. The main reason for the significant ECS3 advantage in area and power is that NST needs CDR to recover data successfully while ECS3 does not. CDR is the main source of power consumption, and even if we use the recently published low-power CDRs proposed in [12, 37, 38, 77], and [73], ECS3 still outperforms NST.

Table 2.4 ECS comparison with simple serial

	Power (μW)			Area (*Gatecount*)			
	SRL[a]	CDR	Total[d] (PI)[e]	SRL	CDR[c]	Total[d] (PI)[e]	
ECS3	19	N/A	19	2098	N/A	2098	65 nm
NST[b]	32.1	70	102.1 (81%)	1327	15,600	16,927 (87%)	90 nm [37]
		62.5	94.6 (80%)		60,000	61,327 (96%)	90 nm [38]
		90	122.1 (84%)		N/A	N/A	90 nm [12]
		57.5	89.6 (79%)		19,800	21,127 (90%)	65 nm [77]
		60.6	92.7 (80%)		N/A	N/A	28 nm [73]

[a] Serializer
[b] NRZ serial transfer
[c] Estimated calculation
[d] SRL+CDR
[e] %Increase as compared to ECS3

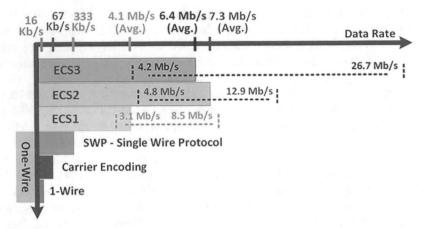

Fig. 2.13 Data rate consumption: one-wire protocols ([16, 27, 44]) vs. ECS1, ECS2, and ECS3

Furthermore, in Fig. 2.13, the ECS3 data rate is compared with the data rates of few existing CDR-less single-wire transmission techniques [16, 27, 44] as well as with ECS1 and ECS2. The comparison shows that ECS3 achieves significantly higher performance without any CDR and with tolerance toward jitters, skew, and clock inaccuracies. For small footprint applications (wireless sensor nodes, wearable computing, body-area networks, etc.) ECS3 is definitely the more reasonable choice.

2.5   Analysis

In this section, we further discuss the major characteristics of the ECS family based on the analytical, numerical, and experimental results we have obtained so far. The detailed timing and robustness analysis is provided in Chap. 3.

2.5.1   Data Rate

ECS is dynamic in that the actual data rate of the protocol is dictated by the pulse count which is very much data dependent. The statistical distributions of the ECS1, ECS2, and ECS3 data rates are shown in Fig. 2.14 for which exhaustive sampling of 16-bit data words is used with a total of $2^{16} - 1$ pseudo-random bit stream (PRBS). Each word is segmented and encoded as per the protocol specifications. For data rate calculations, we use a 25 MHz clock. Please note that the data rates are determined using both numerical simulations and hardware experiments. Comparing the histogram of ECS3 with that of ECS1, an increase in data rate is observed, ranging from 4.2 Mbps (35% ↑) to 26.7 Mbps (214% ↑) with an average

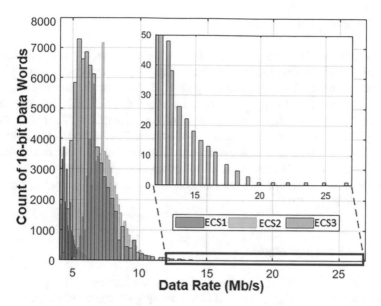

Fig. 2.14 Data rate histograms at 25 MHz clock

of 6.4 Mbps (56% ↑). Additionally, comparing the histogram of ECS3 with that of ECS2, it is observed that ECS3 outperforms ECS2 by achieving a maximum data rate of 26.7 Mbps (107% ↑).

2.5.2 Data Word Length and Complexity

As described in detail in Sect. 2.2.2, the optimum segment length for ECS to maximize data rate is 4 bits. Additionally, in Sect. 2.1, the encoding process generates one flag bit for each of the 4-bit long segments. The four flag bits are then concatenated to form another segment of optimum length (4 bits) that is known as the $CFlags$. Similarly, the process concatenates the $NOIs$ of two consecutive data segments to generate two additional segments of optimum lengths (4 bits each) that are known as the $CNOIs$. By the end of the encoding process, everything is packed in segments of optimum length. Because exactly four 4-bit data segments from the data word are needed to achieve this optimum configuration, the optimum data word size is 16 bits. We note that changing the length of ECS segments to incorporate data words of length larger than 16 bits has the potential of significantly decreasing the data rate while noticeably increasing the complexity, hardware resources, and power consumption.

In this context, ECS3 has a distinct advantage with respect to ECS1 and ECS2 when multiple words are being transmitted. To explain this advantage, we need to take a closer look at two hardware implementations: word-based and block-based.

Word-Based Implementation

The system is strictly designed to handle one word at a time. The transmission starts with setting the busy signal high, as shown in Fig. 2.6, and ends with clearing it when the transmission of a 16-bit word is complete. Let us denote the memory required to store the data word, $CFlags$, and $NOIs$ by M_W, M_F, and M_N bits, respectively. A word-based implementation needs a total of $1 \times M_W + 1 \times M_F + 1 \times M_N$ bits and only one pass to transmit a word, as shown in Fig. 2.15a. On the other hand, transmitting n_W 16-bit words in a word-based implementation incurs an additional delay of one clock cycle for each word to set up the input data port before starting transmission, which results in a total number of $n_W \times (1 + C)$ clock cycles where C is given in Eq. (2.12).

Block-Based Implementation

An alternative system design is to transmit the word block by breaking it into n_W 16-bit words and transmitting each word in a separate pass while keeping a busy signal high unless the transmission of all the words is complete. This is accomplished if the word-based implementation, described above, is used iteratively n_W times with an additional delay of one clock cycle at each pass to set up the transmission of next word in the block. The latency remains $n_W \times (1 + C)$ clock cycles as in the word-based implementation, and the throughput of ECS is unchanged. However, there is an increase in the complexity of the hardware as $n_W \times M_W$ bits of memory are needed to store the full word block. The memories M_F and M_N to store $CFlags$ and $NOIs$ remain unchanged. Additionally, control logic is needed to load the input data port with another 16-bit word from the block and to re-trigger the transmission process, a B_W-to-16 MUX to select a 16-bit word, and an adder to update the selection port of the MUX using a very simple pass control logic. The block diagram of such a block-based implementation is shown in Fig. 2.15b. The added memory and control logic will contribute to a slightly increased consumption of power and hardware resources with the throughput remaining unchanged.

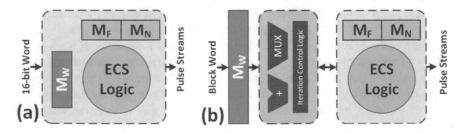

Fig. 2.15 (a) Word-based implementation. (b) Block-based implementation

2.5.3 *Error Detection and Correction*

During ECS transmission, there could be two main sources of transmission errors. These two sources are due to channel noise and are the following:

1. Packet Failure: Either one of the inter-symbol separators or the *CNOIs* is garbled. As shown in Fig. 2.16a-i and a-ii, the receiver *fails to receive* the packet successfully because it expects a number of pulse streams that is different from what is actually transmitted.

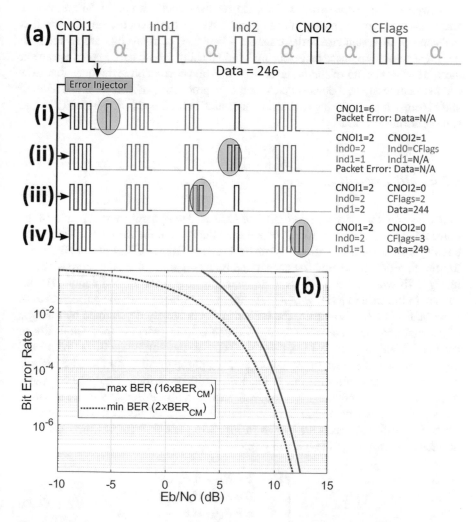

Fig. 2.16 (a) Errors in ECS packet reception (marked in red). (b) ECS BER using BPSK as modulation scheme

2. Data Corruption: The indices or the $CFlags$ pulse streams get garbled due to the addition or removal of pulses. In this scenario, as shown in Fig. 2.16a-iii, the packet *is received successfully*, but the index numbers are wrong, which results in data word errors at decoding. A similar scenario occurs if the $CFlags$ pulses are received with error, as shown in Fig. 2.16a-iv.

In our experimental setup, we have implemented an error injector that randomly inserts extra pulses into the ECS packets, as shown in Fig. 2.16a. The first type of error in the above list can be detected by monitoring the inter-symbol separator. In our implementation of the ECS protocol, if the separator is prolonged for more than twice α (i.e., $separator > 2\alpha$), the receiver declares packet failure, resets, and sends a request to retransmit the packet. As for the second type of error, it is the same as in a standard serial transfer, where one or more bits are in error. In our implementation, these bit errors are handled using a 1-bit parity code. However, there are several state-of-the-art error detection and correction techniques that ECS can use seamlessly as it does not prevent the preprocessing of data prior to encoding and transmission. Exploring such techniques and their compatibility with ECS is the subject of future work.

2.5.4 Bit Error Rate

Like data rate, the bit error rate (BER) of ECS is also dynamic and depends on the transmitted data. The BER of ECS depends on the BER of the PHY layer which may be using a standard modulation scheme [66] such as OOK, ASK, FSK, or BPSK, for transmitting the bit stream. The conceptual block diagram of such a setup is shown in Fig. 2.7b where ECS amounts to an encoding step prior to modulation. The bit stream, in the case of PHY with a modulation, is replaced with the pulse stream as generated by ECS. As a result, the BER of ECS is largely determined by the type of channel and the modulation used. ECS can help in reducing the complexity of the PHY front end by allowing them to focus on the amplitude of the recovered digital signals rather than other factors such as phase or bit width. In case there is a one-bit error in a unit time due to channel modulation, there would be one pulse in error for ECS that could affect the ECS packet in four different locations: within inter-symbol separator, within $CNOI$, within $CFlags$, or within an index number. These four different cases are shown in Fig. 2.16a and will impact the ECS BER differently according to the following rules:

$$BER_{ECS} = \begin{cases} 16 \times BER_{CM} \text{ if } b_e \in \alpha \\ 16 \times BER_{CM} \text{ if } b_e \in CNOI \\ 4 \times BER_{CM} \text{ if } b_e \in CFlags \\ 2 \times BER_{CM} \text{ if } b_e \in index \end{cases} \qquad (2.25)$$

where b_e is the bit in error, BER_{ECS} is the ECS BER, and BER_{CM} is BER of the channel modulation. If the error is within an index number, then an index i is erroneously decoded as index $i - 1$ or $i + 1$. After decoding, two bits may therefore be wrong within a segment: the bit with index i and a bit with index $i - 1$ or $i + 1$. Given that there is a pulse in error, the probabilities of these four cases to occur are 0.5, $1/(n+3)$, $1/2(n+3)$, and $n/2(n+3)$, respectively, where n is the total number of ON bits in all the four encoded segments. With increasing packet size, the error probability decreases when the $CNOI$ and $CFlags$ are corrupted, increases when the indices are corrupted, but remains constant when α is corrupted. Considering BPSK modulation, we have run simulations for the maximum and minimum BER of ECS as shown in Fig. 2.16b. Regardless of the incoming data, the ECS BER will always fall between these two extrema.

2.5.5 Pulse Width and Shape

The pulse *width* is not a primary ECS parameter because the reception technique exploits the detection of *edges* in contrast with standard serial bit transfer, where the pulse width (inverse of the baud rate) is a primary parameter. ECS is therefore capable of working with a non-ideal pulse waveform, thus enabling great flexibility in pulse shaping. All pulse shapes are allowed as long as they satisfy switching and peak detection constraints and do not overlap with each other. A few examples are shown in Fig. 2.17.

2.5.6 Reliability

The reliability of ECS is an indication of the likelihood of successful transmission. ECS reliability is measured in terms of the *number of vulnerable locations (NVLs)*. An *NVL* is a sub-segment within the ECS packet which, if corrupted, can cause a

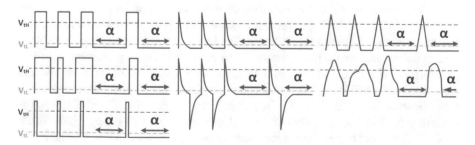

Fig. 2.17 Examples of valid signals for ECS transmission. V_{tH}: high signal level threshold. V_{tL}: low signal level threshold

packet failure. The inter-symbol separators α and the $NOIs$ are examples of $NVLs$ in ECS packet. Indeed, the longer α is, the higher the corruption likelihood, and therefore, the more likely packet failure is. Similarly, a corrupted NOI results in packet failure, and the failure likelihood increases with the number of transmitted $NOIs$. The packet failure due to these $NVLs$ has already been discussed in Sect. 2.5.3. In the ECS family of protocols, ECS1 has $NVLs$ in the range of 10–18. On the other hand, the $NVLs$ of ECS2 are reduced to a fixed number of 6, and as a result, ECS2 improves transmission reliability by about 66.7% with respect to ECS1. As for ECS, the compactness of its transmission packet restricts the $NVLs$ to the range of 5 to 12, which results in a reliability improvement of up to 72.2% with respect to ECS1 and makes it competitive with ECS2.

2.5.7 Robustness

As highlighted in Sect. 2.5.5, ECS is indifferent to pulse shape and width because only edges are used for decoding at reception. This important ECS property results in a remarkable tolerance toward clock discrepancies, jitters, and skews. A detailed analysis of these clocking and timing aspects are presented in Chap. 3. Therein, a threshold on clock rate difference between transceivers is derived, below which the protocol operates with zero decoding error over an ideal channel. It must be noted that this threshold does not define a boundary beyond which ECS needs a synchronization mechanism. It is rather an indicator when ECS needs to change its protocol parameters to enable communication with network nodes that are operating at significantly different clock rates. Chapter 6 highlights the methodology to establish a successful communication link automatically when the abovementioned threshold on clock difference is crossed. In the ideal scenario where there are no clock skews or jitters, the threshold is determined by α and α_{th}. For the minimum recommended settings of $\alpha = 4$ and $\alpha_{th} = 2$, the upper bound on clock rate difference for a 25 MHz transceiver is 10.3 MHz. Beyond this difference, it is impossible for the receiver counter on the slow transceiver to reach the threshold value of 3 or higher, as discussed in Sect. 2.5 and illustrated in Fig. 2.6. Clock jitters and skews result in the decrease of this ideal upper bound on clock rate difference. The possible ranges for clock skews and jitters are presented in Chap. 3, where a trade-off between reliability, defined as robustness with respect to clock rate variations, and data rate is quantified and used for protocol parameter selection. Using $\alpha = 4$, an ECS clock equal to the system clock of 25 MHz, and worst-case clock skew and jitter, the upper bound on clock rate difference between ECS transceivers is 4.6 MHz ($\sim$20%). In practice, this means that if one ECS end is operating at 25 MHz, the other ECS end can operate in the range [20.4 MHz, 29.6 MHz] without impacting the reliability of the transmission. One important area of future investigation is the impact of non-ideal transmission channels on ECS robustness.

2.5.8 Overall Latency

The impact of ECS on the latency of the communication system very much depends on the ECS hardware implementation. The minimum latency is one clock cycle, and to achieve it, the four data segments can be encoded together before starting transmission. The drawback of this approach is that the encoder hardware and the memory used to store encoding information would be quadrupled. The alternative that we have adopted is a pipelined datapath where only two segments are encoded before starting transmission. Indeed, at least two segments need to be encoded to generate $CNOIs$ and initiate the transmission process, as explained in Sect. 2.1.3. This pipelined architecture requires only twice the encoding and memory hardware. The same hardware is reused to encode the two segments, while the indices of the previous two segments are being transmitted. The encoder hardware is combinational, which results in a latency of one clock cycle only. On the other hand, sequential implementation of the encoding process would introduce a latency of more than one clock cycle.

2.5.9 Networking

ECS is architecturally flexible in that it supports a wide variety of networking options. It can be configured in various network topologies, including Master–Slave, Star, Ring, Tree, and Peer-to-Peer. In a single-channel implementation (Chap. 9), several MSP430 cores have been used to establish a Master–Slave network of low-end devices, as shown in Fig. 2.18a. The same setup can be used to establish a ring network, as shown in Fig. 2.18b. In the latter, where communication is restricted to nearest-neighbor devices, device IDs have been used to enforce the ring topology.

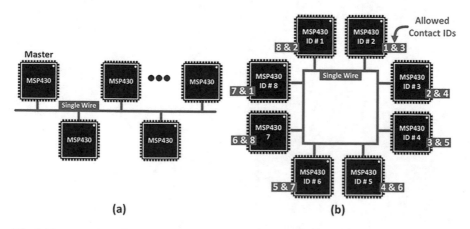

Fig. 2.18 Network topology [50]: (**a**) Master–Slave, (**b**) configuring ring

For example, MSP430 with ID number 2 can only communicate with the devices having ID numbers 1 and 3. Device IDs can similarly be used to implement a network of ECS devices without any change in the underlying hardware.

2.6 Conclusions

In conclusion, ECS is a novel, single-channel, communication protocol that simultaneously meets the requirements of low power, high data rate, reliability, and secure transmission for device-to-device communication between constrained edge nodes. ECS reduces silicon area and power consumption significantly by eliminating the need of power- and area-hungry circuits for clock and data recovery. This is because ECS packet reception and decoding are based on counting the rising edges of the transmitted pulses which makes the pulse width inconsequential. ECS is robust with respect to skews, jitters, and clock variations and combines the best features of both ECS1 and ECS2 with respect to data rate, reliability, packet security, and power efficiency. In summary, ECS is the better choice for constrained devices in a variety of use cases, including small footprint transceivers, wireless sensor nodes, implantable devices, and body-area networks. The protocol can be applied to other communication media such as photonics, infrared, and visible light.

Chapter 3
Timing and Robustness Analysis

Due to the lack of secured timing sources globally available today, a reasonable approach to securing time is to ensure systems can maintain timing within the tolerance of their application for the duration of a timing compromise.

Cyber-Physical Systems Public Working Group, NIST

The objective of this chapter is to provide a full theoretical framework that underlies the timing and robustness of ECS protocols. This framework is used to derive rigorously the performance parameters that we discovered empirically in Chap. 2, including those related to ECS robustness with respect to clocking uncertainties. The most important result of the ECS theory presented in this chapter is the crucial role played by the delay (expressed in transmitter clock cycles) between data segments in a given packet. For a given encoding scheme, this delay determines both the ECS average data rate and the maximum clock uncertainty tolerance. Rigorous timing and robustness analysis is provided to quantify the ECS robustness margin in the presence of transmitter-to-receiver clock variations as well as clock skew and jitter within each clock. To conduct the analysis, we have found it convenient to select the ECS1 packet format of Fig. 2.9.

3.1 Timing and Robustness Analysis

Robustness is directly related to the capability of behaving appropriately in the presence of different sources of errors. In this section, we first survey the sources of errors in ECS1 channels. Then we describe the intrinsic timing parameters of the ECS1 protocols and introduce some of the constraints they must satisfy for error-free operation. The last subsection is devoted to deriving a closed-form upper bound on clock discrepancy between transmitter and receiver for error-free operation in the presence of clock variations.

© Springer Nature Switzerland AG 2022   37
S. Muzaffar, I. M. Elfadel, *Secure, Low-Power IoT Communication Using Edge-Coded Signaling*, https://doi.org/10.1007/978-3-030-95914-2_3

3.1.1   Sources of Errors

Clock discrepancy between transmitter and receiver is one of the main sources of errors and creates significant trouble in digital communication systems. The problem becomes even more serious when single-channel communication is used. The clocks of both ends need to be synchronized to limit the errors for which a variety of techniques such as clock-and-data recovery (CDR) circuits are available. ECS1 is an ultra-low-power single-channel protocol, without CDR, but with unique robustness properties with respect to clock variations. Such variations are expected to be very wide-spread in the IoT environment where IoT devices with different clocking and performance requirements need to communicate. Each end of the ECS1 communication link is comprised of a transmitter and a receiver. The subsequent formulation and calculations are carried out assuming the transmitter is running at a slow clock rate f_S while the receiver is running at fast clock rate f_F. The link setup is shown in Fig. 3.1a and b. In the remainder of this chapter, we adopt

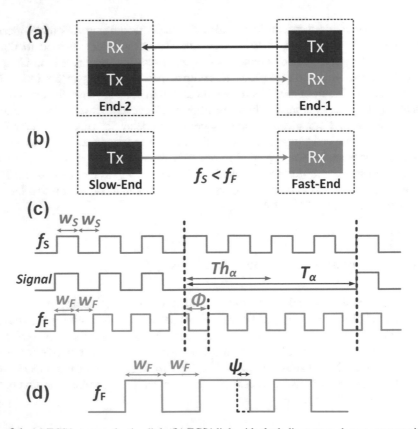

Fig. 3.1 (**a**) ECS1 communication link. (**b**) ECS1 link with clock discrepancy between transmitter and receiver. (**c**) ECS1 waveforms with clock discrepancy. (**d**) Receiver clock jitter

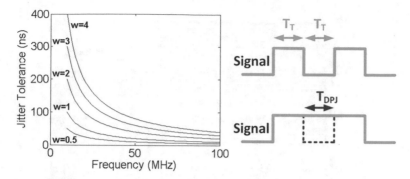

Fig. 3.2 Effect of frequency on jitter tolerance

the convention that all the parameters with S subscript are for the slow end and with F subscript are for the fast end.

The sources of timing errors in ECS1 are as follows:

1. The data pulse jitter is the main source of time difference between two adjacent data pulses. If such jitter is high, the pulses start overlapping, which introduces a missing-pulse error at the receiver. The extent to which ECS1 can tolerate these jitters is explained using Fig. 3.2, where T_T is the transmitter clock time period. The faster the clock, the closer the data pulses are to each other, and therefore a small jitter may lead to successive pulse overlap. This explains the plot in Fig. 3.2, which shows that data pulse jitter tolerance is inversely proportional to transmitter clock rate. For a given clock rate, the data pulse jitter tolerance increases with the increase in pulse width coefficient w.

2. The phase shift Φ is the time difference between the edge of the receiver clock, f_F, and the transmitter clock edge marking the start of the inter-symbol delay, T_α. The phase shift may affect the detection of inter-symbol delay especially in the presence of clock discrepancies. Phase shift ranges from 0 to T_F, the period of the receiver clock, as shown in Fig. 3.1c. In terms of shift percentage φ, we have $\Phi = \varphi T_F, 0 \leq \varphi \leq 1$.

3. Receiver clock jitter, Ψ, may also affect the detection of inter-symbol delay. The jitter value is related to the receiver clock period T_F by $\Psi = \psi T_F, 0 \leq \psi \leq 0.5$, where ψ represents jitter percentage. Receiver clock jitter is shown in Fig. 3.1d.

4. Noise associated with the off-chip environment can have an effect on data pulses. The increase or decrease in pulse levels, due to the external noise, makes it difficult for the receiver to detect pulses correctly. Depending on the noise level, an extra pulse may be detected or a pulse may be skipped. In both cases, one gets a decoding error. To analyze the performance of ECS1 in the presence of noise, the encoded pulse stream of data is exposed to white Gaussian noise. The noisy signal is filtered at the receiver end, then decoded according to the ECS1 algorithm, and the number of errors encountered is counted. The results are plotted in Fig. 3.3 for different values of E_b/N_0 (the ratio of energy-per-bit

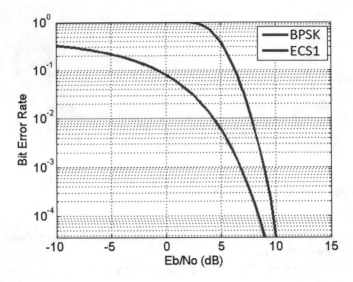

Fig. 3.3 BER analysis

to noise power spectral density) for both BPSK and ECS1. ECS1 is less immune
to noise as compared to BPSK, but its BER rapidly reduces to zero at E_b/N_0
of $\sim$10.5 dB. For the lower signal-to-noise ratios, a lightweight error correction
scheme can improve the BER albeit at the expense of a small increase in power
consumption. A full system-level analysis of this trade-off is the subject of on-
going work. In this chapter, we assume that the BER is zero.

3.1.2 Pulse Width Coefficient

The data pulse width coefficient, w, is the number of transmitter clock cycles during
which the pulse remains high. An example is shown in Fig. 3.1c. The inter-symbol
separation increases linearly with the pulse width as will be explained in the next
subsection. On the other hand, the data rate reduces with inter-symbol separation.
For a 50% duty cycle, the total pulse duration $\Pi = 2wT$, where T is the clock
period of the transmitter, see Fig. 3.1c. A pulse width equal to half-clock cycle (i.e.,
$w = 0.5$) results in the highest data rate and lowest requirement on the inter-symbol
separation coefficient (see next paragraph). With the same pulse widths, for both the
transmitter and receiver (i.e., $w_S = w_F$), the transmitted pulse time period would
be equal to the time period at receiver (i.e., $T_S = T_F$). However, in the presence of
clock discrepancies; $w_S \neq w_F$, the transmitted data pulse with time period T_S will
not map exactly to the time period T_F at the receiver. Instead, we have

$$T_S = \frac{w_S}{w_F} T_F \implies w_S = \frac{f_F}{f_S} w_F. \tag{3.1}$$

This relationship expresses the invariance of the *number* of transmitted pulses. This number depends solely on the product of the pulse width and the corresponding clock rate.

3.1.3   Inter-symbol Separation Coefficient

The inter-symbol separation coefficient, α, is the number of transmitter clock cycles used to separate the pulse streams of two adjacent symbols. This is the most important parameter in ECS1 design as it affects both its data rate and reliability. A badly selected α will result in increased decoding errors and possibly complete failure. As will be seen in the following paragraphs, an efficient ECS1 protocol requires that α be an even number. The smallest possible α value is 4. $T_{\alpha s}$, the inter-symbol separator in terms of transmitter time period (T_S) is shown in Fig. 3.1c and is calculated as $T_{\alpha s} = \alpha_S T_S$. To generate such time interval, the transmitter pulls the line low and keeps it in that state for α clock cycles. For a successful reception, the inter-symbol interval at both ends must satisfy the *time* invariance principle:

$$\alpha_F T_F = \alpha_S T_S \tag{3.2}$$

In the presence of clock discrepancies, phase shift, and clock jitter, the time invariance condition is expressed as

$$\alpha_F T_F = \alpha_S T_S + \varphi T_S + \psi T_S \implies \alpha_S = \frac{f_S}{f_F}\alpha_F - (\varphi + \psi) \tag{3.3}$$

The correct ECS1 transmission and reception of data depend on several parameters that need to be selected judiciously. These parameters include the delay threshold and the inter-symbol interval.

Inter-symbol Interval Threshold

A portion of inter-symbol interval, called the interval threshold Th_α, shown in Fig. 3.1c, is used at the receiver to discriminate between data pulses and inter-symbol separator. The optimal threshold is given by

$$Th_\alpha = \frac{T_\alpha}{2} = \frac{\alpha_F T_F}{2} = \alpha_{Th} T_F \tag{3.4}$$

where $\alpha_{Th} = \alpha_F/2$ is the separation threshold coefficient. Th_α ensures that the receiver clock-cycle count does not decrease during the reception of the inter-symbol separator to the extent that the receiver stops detecting it as inter-symbol separation. Also, the cycle count should not increase to the extent that the receiver starts detecting data pulses as inter-symbol separator. The absence of such optimal threshold will result in decoding errors due to either pulse undercounting or inter-symbol interference.

Selection of Inter-symbol Separation Coefficient

To distinguish the data pulses and the inter-symbol separators, the transmitter-generated delay should be longer than the duration of one data pulse. The inter-symbol delay coefficient should therefore satisfy

$$\alpha_S T_S > T_S + \varphi T_S + \psi T_F \implies \alpha_S > 1 + \varphi + \psi \frac{f_S}{f_F} \tag{3.5}$$

where local clock discrepancies due to receiver phase shift and clock jitter are accounted for. Using the maximum possible values of 1 and 0.5 for φ and ψ, respectively, in (3.5) and assuming $f_F = f_S$, we get $\alpha_S > 2.5$.

In theory, the integer α_S can be chosen equal to 3, but from a hardware implementation view point we set it equal to 4 as it is easier to implement multiplication and division of power 2 numbers using left or right shift operations. In the next subsection, we study the interplay between inter-symbol separator and robustness with respect to clock variations.

3.1.4   Clock Discrepancy Tolerance

Given the slow-end clock frequency, f_S, and the parameters $\alpha_S, \alpha_F, w_S, w_F, \psi$, *and* φ of the ECS1 system, our goal now is to find the highest possible fast-end clock frequency f_{Fmax} above which decoding errors start to occur. To find f_{Fmax}, the condition for error-free operation should be fulfilled, namely, the pulse duration should be less than the inter-symbol threshold

$$2w_S T_S \le \frac{\alpha_S}{2} T_S \tag{3.6}$$

Using w_S from (3.1) and α_S from (3.3), we get

$$4w_F f_F^2 + (\varphi + \psi) f_S f_F - f_S^2 \alpha_F \le 0 \tag{3.7}$$

which is satisfied if and only if

$$\beta \equiv \frac{f_F}{f_S} \le \left[\frac{\sqrt{(\varphi + \psi)^2 + 16 w_F \alpha_F} - (\varphi + \psi)}{8 w_F} \right] \equiv \beta_{max} \tag{3.8}$$

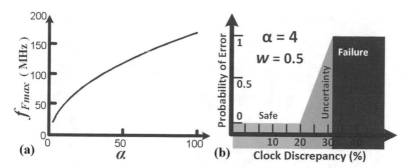

Fig. 3.4 (a) Discrepancy limit vs. inter-symbol separation α. (b) Regions of operation ($f_S = 25\,\text{MHz}$)

which is the main theoretical result of this chapter. If $\beta \leq \beta_{max}$, ECS1 transmission will be error free. It is interesting to note that β_{max} is linear in the receiver clock jitter and phase shift but varies as the square root of receiver inter-symbol separation.

If we are to keep the inter-symbol delay coefficient α the same for both ends, the rate of the fast clock should not exceed the limit imposed by $f_{Fmax} = \beta_{max} f_S$. In Fig. 3.4a, we plot f_{Fmax} for $f_S = 25\,\text{MHz}$. Of course, for error-free transmission, $f_F \leq f_{Fmax}$. Figure 3.4b identifies different regions of operation. The safe region of operation is marked up to the limit calculated using f_{Fmax}. Beyond this, there is a region of uncertainty in which errors start occurring randomly. At a certain level of clock discrepancy, total failure occurs due to the failure in detecting even a single inter-symbol separator. The recommended region of operation is of course the one delimited by f_{Fmax}.

3.1.5 Selection of Inter-symbol Separation Coefficient

Another interpretation of the inequality (3.7) is as a design formula for selecting an appropriate value of receiver inter-symbol separation in the presence of clock discrepancies between transmitter and receiver and the presence of local clock non-idealities at the receiver. Solving for α_F in (3.7), we get

$$\alpha_F \geq \lceil 4\, w_F \beta^2 + \beta\,(\varphi + \psi) \rceil \equiv A_F \tag{3.9}$$

After selecting an inter-symbol separation coefficient for the slow end α_S, the inter-symbol separation coefficient for the fast end α_F is set equal to α_S if $f_F \leq f_{Fmax}$. Otherwise, α_F is a scaled version of α_S given by

$$\alpha_F = mod(A_F, 2) + A_F \tag{3.10}$$

This formula guarantees that α_F is the smallest even integer satisfying (3.7).

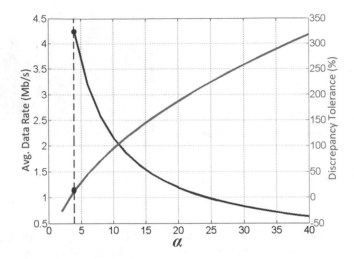

Fig. 3.5 Trade-off between robustness and data rate

3.1.6   Summary on Inter-symbol Separation

The larger α, the more resilient ECS1 is to timing variability. This however is achieved at the expense of significantly reduced data rates and increased power consumption. This trade-off is illustrated in Fig. 3.5. A careful adjustment of α is necessary to meet the requirements of a particular application. The α value used in all our experiments is 4, with $w = 0.5$. Values of α less than 4 will result in transmission failure *even* if the transmitter and receiver are running on ideal clocks with the same frequency. Therefore, the minimum allowable value of α is 4.

3.2   Protocol Failure Modes and Error Correction

ECS1 transmits all the information in the form of a pulse train. A common case of erroneous transmission happens when there is a pulse miscount. In this case, ECS1 stops reception immediately and keeps it in a "halt" state until an explicit reset signal is sent, at which time transmission is resumed. With the use of a simple counter, the halt state provides a useful error detection mechanism. The counter is activated via the receiver's busy signal, which remains active when the error occurs. The counter gets reset at each falling edge and continues to count until the next falling edge. If the count reaches a threshold value, Th_{Error}, the state is considered an error and thus a reset signal is generated. An appropriate value of the error detection threshold is

$$Th_{Error} = 2(\alpha + 2\,w\,l) \tag{3.11}$$

which is twice the number clock cycles needed for the inter-symbol separation.

A variety of available techniques [65] can be employed to handle an error condition. One simple recommendation is to request the transmitter to resend the data. Another recommendation is for the configurable layer of the protocol to send an acknowledgment for each of the successful transfers. The only error that cannot be detected using the count method is the distortion of middle pulses in the index pulse train. In such cases, there will be no halt state and the receiver will infer a wrong index number. This transmission error can be handled using a simple parity check or other similar methods. However, such pulse distortion error is very unlikely as it occurs only in the presence of excessive external noise.

3.3 Experimental Verification

An experimental setup comprised of two IoT nodes communicating using ECS1, as shown in Fig. 3.6, is used to verify the limitations imposed by (3.8) on the maximum tolerable clock frequency for the fast end. Each node is comprised of an ECS1 protocol module (SED), a Logical Topology Control module (LTC), a PHY layer, and a Test Runner, as shown in Fig. 3.6a. The SED and PHY are implemented in Verilog HDL. The LTC and test runner are implemented using the Verilog IP of TI MSP430 microcontroller. The whole setup is implemented in Verilog on the Xilinx Virtex-7 FPGA platform. Two clocks, one for each node, are generated with the help of a Virtex-7 on-chip PLL, as shown in Fig. 3.6b. The slow-node clock is fixed at 25-MHz, but the rate of the fast-node clock is increased gradually from 25-MHz. Using $\alpha = 4$, $w = 0.5$, $\varphi = 1$, and $\psi = 0.01$ at both ends, we have $\beta_{max} = 1.2$. The LTC of the slow node directs the ECS1 transmitter to send the 16-bit data starting at 0 with an increment of 1 at each transmission. The fast end receives the data and replies back the same. The returned and original data words are compared to verify the complete round-trip chain. The experiment confirmed the results of (3.8) that the ECS1 transmission works flawlessly until the clock frequency of the fast node reaches $\approx$30-MHz (20% discrepancy), which is in agreement with the theoretical bound of (3.7).

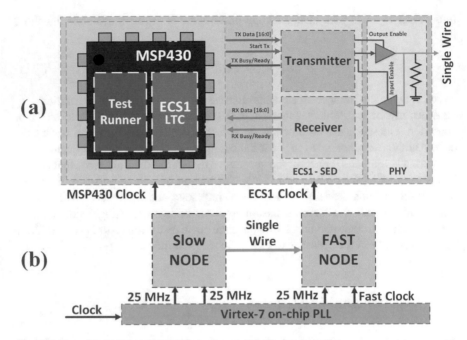

Fig. 3.6 Experimental setup (**a**) Inside each node (**b**) Clocks generation

3.4 Conclusions

In this chapter, we have presented a detailed timing analysis for ECS protocols to meet robustness requirements. In the analysis, the inter-symbol separation parameter used to delimit the boundary between data segments in a packet plays a significant role. In addition to the gained insight into ECS1, the analysis helps quantifying the trade-off between data rate and robustness. Based on our analysis, an inter-symbol separation coefficient α of 4 clock cycles and a pulse width w of half-clock cycle are recommended. These recommended design parameters have been experimentally verified using a Xilinx Virtex7 FPGA platform that illustrates the simplicity, efficiency, and reliability of using ECS as a single-channel communication protocol for IoT devices. These additional ECS features augment the already proven ones of low-power, high data rate, and small footprint.

Chapter 4
Doubling the ECS Data Rate

Better three hours too soon than a minute too late.

William Shakespeare

The data rate of these edge-coded schemes is a function of clock frequency because at least one clock cycle is needed to generate one pulse. One possible method to achieve higher data rates is to increase clock frequency, but in the context of constrained IoT edge nodes, such increase will result in high power consumption and more complex circuitry to implement frequency scaling. Another limitation on edge nodes is the limited range of frequency scaling and the upper bound that the node may impose on the available clock rates. In this chapter, we propose an alternative method to improve the data rate using edge coding based on both the rising and falling edges of the transmitted pulses. The contrast between the original single-edge-coding scheme and the proposed double-edge-coding one is illustrated in Fig. 4.1. Intuitively, the double-edge scheme should achieve twice the data rate of the single-edge scheme. However, the hardware implementation of the latter coding is not trivial and presents several design challenges. In this chapter, we use the single-edge coding of edge-coded signaling (ECS) as a starting point and present an efficient hardware implementation of the double data rate (DDR) signaling transceiver that remains within the power and area envelop of the single-edge scheme. The proposed hardware design of DDR-ECS is implemented over an FPGA platform and is further synthesized using GLOBALFOUNDRIES $65nm$ CMOS technology. The synthesis results show that the proposed coding scheme indeed achieves double data rate while remaining well within the power and area budgets of the standard edge-coded schemes.

© Springer Nature Switzerland AG 2022
S. Muzaffar, I. M. Elfadel, *Secure, Low-Power IoT Communication Using Edge-Coded Signaling*, https://doi.org/10.1007/978-3-030-95914-2_4

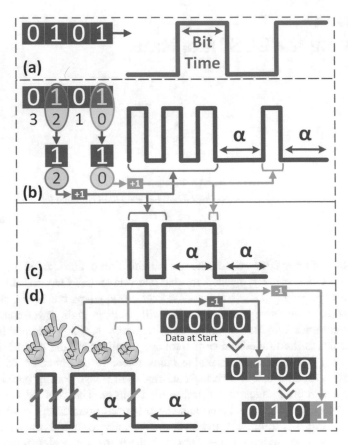

Fig. 4.1 (**a**) Standard serial transfer. (**b**) Single-edge-coded signaling. (**c**) Double-edge-coded signaling. (**d**) Double-edge-coded reception

4.1   Single-Edge Scheme: An Example

Let us consider an example of ECS's single-edge scheme to prepare a stage to understand the double-edge scheme better. Assume we want to transmit a decimal number "267" that is "0000 0001 0000 1011″ in binary representation. First, ECS breaks the B-bit data word into segments of size l and generates $N = B/l$ segments. In our example, we have used $B = 16$ and $l = 4$, which results in four 4-bit segments $S_1 = 1011$, $S_2 = 0000$, $S_3 = 0001$, and $S_4 = 0000$. Second, each segment is encoded. The ECS encoding process inverts the segment bits if the number of ON bits is greater than half the segment size (i.e., $>l/2$). A Flag bit for each segment is used to indicate if the segment is encoded or not. All the four segments are concatenated to generate a combined 4-bit flag, $CFlags =$

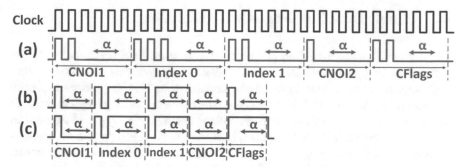

Fig. 4.2 (**a**) Standard ECS transmission (Data $= 267$). (**b**) DDR-ECS transmission (Data $= 267$). (**c**) DDR-ECS transmission (Data $= 132$)

$\{Flag4, Flag3, Flag2, Flag1\}$. A 2-bit NOI flag is used to specify the number of ON bits in an encoded segment. The $NOIs$ of two consecutive segments are concatenated to generate two combined 4-bit $NOIs$, $CNOI1 = \{NOI2, NOI1\}$ and $CNOI2 = \{NOI4, NOI3\}$. In our example, only segment S_1 satisfies the encoding condition $(3 > l/2)$ and, therefore, the bits are inverted to yield an updated segment $S_1 = 0100$ and the respective flag bit, $Flag1$, is set high to indicate the inversion. At the end of the encoding process, the combined flags would be $CFlags = 0001$. The $NOIs$ would be $NOI1 = 1$, $NOI2 = 0$, $NOI3 = 1$, and $NOI4 = 0$; hence, $CNOI1 = 0001$ and $CNOI2 = 0001$. Next, the ECS transmitter selects the ON bits only in each encoded segment. In our example, we have index number 2 from S_1 and index number 0 from S_3. At this stage, all the information pieces are packeted together to construct a packet. Each piece in the packet is then transmitted in the form of a pulse stream where the number of pulses is equal in count to the number represented by the piece, as shown in Fig. 4.2a. Please note that ECS sends an additional pulse for each pulse stream (e.g., $CNOI + 1$ or $index + 1$). For indices, the additional pulse is necessary to handle the encoding of index 0. Otherwise, no pulse will be transferred if the bit at index 0 is ON. For $CNOIs$, the additional pulse is needed to inform the receiver not to expect any index number when the count of ON bits in the first two segments is zero. For $CFlags$, an additional pulse is needed to represent zero content of $CFlags$ as in the case when no segment goes through the encoding inversion. All the pulse streams are separated by an optimum inter-symbol separation of $\alpha = 4$ clock cycles, as shown in Fig. 4.2a. At the receiver end, the rising or falling edges of all the pulse streams are counted to infer all the pieces in the packet. The received information is then used in the decoding process to rebuild the transmitted data. The ECS is dynamic in that even two consecutive data words may result in a different number of pulses and, therefore, different data rates.

4.2   Double Data Rate Edge-Coded Signaling

The transmission and reception process of the standard ECS, as described in the previous section, employs only one pulse edge in counting pulses, while the other edge remains unused. The double data rate ECS (DDR-ECS) puts these unused edges to work. Indeed, DDR-ECS uses both the rising and falling edges to transmit an edge stream instead of transmitting a pulse stream. For example, if we need to transmit "4," the ECS needs four pulses identified with their rising edges, whereas DDR-ECS needs two rising and two falling edges. Let us take a concrete example. To transmit a number "267" using DDR-ECS, the packet formation process is the same as for ECS. However, the transmission process will differ as per the following rules:

1. The transmission always starts with a low signal level.
2. Toggle the signal level at each iteration of the edge-stream counter.
3. The counter should increment at both pulse edges. If the count of edges is odd, the final increment will use only one pulse edge.
4. Keep the last signal level during the inter-symbol interval.
5. The counter for the inter-symbol interval should also follow Rule 3.
6. At the end of the packet (e.g., CFlags), force the signal low at the end of the inter-symbol interval.

The two DDR-ECS transmission examples are shown in Fig. 4.2b, c. It is observed that the transmission time in DDR-ECS is half of the single-edge one, and therefore, the data rate is doubled. The next section presents a small footprint, high-performance hardware implementation of DDR-ECS.

4.3   Hardware Implementation

4.3.1   Transmitter

The hardware for the DDR-ECS transmitter comprises three main modules: edge-stream transmitter, encoder, and a finite state machine, as shown in Fig. 4.3a. The edge-stream transmitter inputs a count of edges and transmits an edge stream along with an inter-symbol separator. The module starts sending edges when the *Start Edge Stream* signal is set high by the FSM, and sends the *End-of-Separator* signal to the FSM when the transmission of an edge stream is complete. If the module is active, either a positive or a negative edge counter is incremented at each edge of the clock, and the output is toggled. The process continues unless the total count of edges reaches the input edge-stream count. The encoder inputs a 4-bit segment, applies ECS encoding, and outputs the number of ON bits, index numbers, and the corresponding flag bits. The encoder is combinatorial with its truth table given in Fig. 4.3e. Because we need to generate two $NOIs$ to form one $CNOI$, two

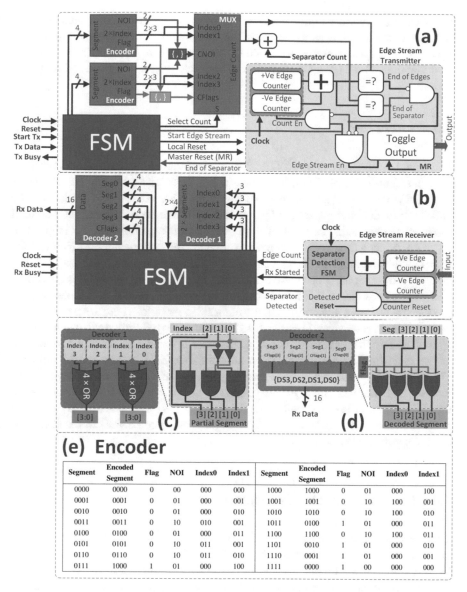

(e) Encoder

Segment	Encoded Segment	Flag	NOI	Index0	Index1	Segment	Encoded Segment	Flag	NOI	Index0	Index1
0000	0000	0	00	000	000	1000	1000	0	01	000	100
0001	0001	0	01	000	001	1001	1001	0	10	100	001
0010	0010	0	01	000	010	1010	1010	0	10	100	010
0011	0011	0	10	010	001	1011	0100	1	01	000	011
0100	0100	0	01	000	011	1100	1100	0	10	100	011
0101	0101	0	10	011	001	1101	0010	1	01	000	010
0110	0110	0	10	011	010	1110	0001	1	01	000	001
0111	1000	1	01	000	100	1111	0000	1	00	000	000

Fig. 4.3 Hardware block diagram. (**a**) Transmitter. (**b**) Receiver. (**c**) Receiver decoder 1. (**d**) Receiver decoder 2. (**e**) Transmitter encoder

encoders are integrated in parallel. All the encoder outputs are routed to the edge-stream transmitter via a MUX that forwards a piece of a packet for transmission with the selection bits based on inputs from the FSM. The FSM is the main control logic of the transmitter and maintains the desired flow of the modules. The FSM inputs a 16-bit data to transmit, asks encoders to encode the segments, iteratively directs the MUX to forward an edge count to the edge-stream transmitter, and for each forwarded count, activates the edge-stream transmitter. At the completion of each edge stream, a *Local Reset* is generated to make the stream transmitter ready for the next iteration. The DDR-ECS transmitter FSM state diagram is given in Fig. 4.4a.

4.3.2 Receiver

The main building blocks of a DDR-ECS receiver include an edge-stream receiver, decoders, and a control FSM, as shown in Fig. 4.3b. The edge-stream receiver is activated when a rising edge of an input edge stream has arrived. The module counts both the positive and negative edges of the input stream unless the inter-symbol separator is detected. The FSM of the inter-symbol separator detector is given in Fig. 4.4c. At each rising edge of the receiver clock, the current and the previous total edge counts are compared. If the counts match, it is an expectation of the arrival of inter-symbol separator since there is no input edge following the last FSM iteration. The FSM enters the verification mode, and if the edge count equality persists, the separator is declared and notified to the main FSM module. There are two decoders in the DDR-ECS receiver. Decoder 1 is responsible for rebuilding the two segments by looking into the input index numbers. Decoder 2 decodes all the four segments as per the received $CFlags$. The decoders are combinatorial in nature and are shown in Fig. 4.3c, d. The main FSM gets activated at *Rx Started* signal from the edge-stream receiver and waits for the *Separator Detected* signal. At each separator detection, the FSM collects the edge count, puts the count at the target position in the packet frame, and asks the decoders to rebuild the data word. The DDR-ECS receiver FSM state diagram is given in Fig. 4.4b.

4.4 Formulation and Optimizations

This section analyzes the effect of the proposed double-edge-coding scheme on protocol performance as compared to the standard ECS. Let b_i^s be the i-th bit, NOI_s be the number of ON bit indices, and F_s be the flag bit for the s-th encoded segment. Also, let C_s be the number of clock cycles required to generate the edge streams for the encoded segment indices, and let CI_x and CF_y be the concatenated number of indices $CNOI$, and the concatenated flag bits $CFlags$, respectively. Furthermore, denote by C_α the number of clock cycles per inter-symbol separation, by C_t the total number of clock cycles needed to transmit an ECS packet, by T the clock period, and by R the data rate. The mathematical expressions linking these quantities for DDR-ECS are given in Table 4.1(a). They are similar to those of the standard ECS [56] except that the number of clock cycles is approximately half. Remember that both the rising and falling edges are being used in DDR-ECS to transmit the edge

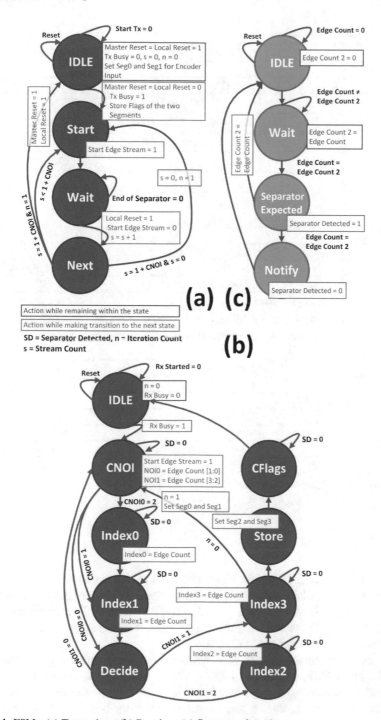

Fig. 4.4 FSMs. (**a**) Transmitter. (**b**) Receiver. (**c**) Separator detection

Table 4.1 Comparison and results

		ECS (Chap. 2 [56])	DDR-ECS
(a) Formulation			
1	R	$B/(TC_t)$	
2	C_t	$C_\alpha + \sum_{x=1}^{N/2} CI_x + \sum_{s=1}^{N} C_s + \sum_{y=1}^{N/l} CF_y$	
3	C_α	$(N/l + N/2 + \sum_{s=1}^{N} NOI_s)\alpha$	$ECS\_C_\alpha/2^{\text{a}}$
4	CI_x	$1 + NOI_{2x-1} + 2^{l/2}NOI_{2x}$	$\lceil ECS\_CI_x/2 \rceil$
5	NOI_s	$\sum_{i=0}^{l-1} b_i^s$	ECS_NOI_s
6	C_s	$\sum_{i=0}^{l-1}(i+1)b_i^s$	$\sum_{i=0}^{l-1} \lceil (i+1)b_i^s/2 \rceil$
7	CF_y	$1 + \sum_{i=0}^{l-1} 2^i F_s$ $s = i + l(y-1)$	$\lceil ECS\_CF_y/2 \rceil$
(b) Optimization			
8	$E[CI_x]$	$(2 + l(1 + 2^{l/2}))/2$	$ECS\_E[CI_x]/2$
9	$E[CF_y]$	$(1 + 2^l)/2$	$ECS\_E[CF_y]/2$
10	$E[C_s]$	$l(l+1)/4$	$l(l+2)/8$
11		$f(l) \triangleq \partial E[C_t]/\partial l = 0$ $(8 + 2l)\alpha + 2l - l^3(2^{l/2}\ln(2) + 1) - 2^l(2l\ln(2) - 4) + 4 = 0$ (ECS) $(32 + 8l)\alpha + 4l - l^3(2^{l/2}\ln(2) + 2) - 2^l(4l\ln(2) - 8) + 8 = 0$ (DDR-ECS)	
12	l_{opt}	$2.83 \approx 4$ bits	$3.16 \approx 4$ bits
13	α_{opt}	4 clock cycles	2 clock cycles
(c) Performance (25 MHz clock)			
14	Data rate[b]	4.2–26.7 (6.4 Avg.)	7.8–44.4 (12 Avg.)
15	Power[c]	$\approx$19	$\approx$19
16	Eb[d]	0.7–4.5 (2.9 Avg.)	0.4–2.4 (1.6 Avg.)
17	Area[e]	$\approx$2098	$\approx$1943

[a]The notation $ECS\_$ represents the equation from the ECS column
[b]Mb/s
[c]μW
[d]pJ/bit
[e]Gate count

streams. We need exactly half the number of clock cycles to transmit an even number of edges (e.g., 3 cycles for 6 edges), whereas if the number of edges is odd, we need a full clock cycle for the last edge (e.g., 3 cycles for 5 edges). To incorporate the odd number of edges in the mathematical expressions, the ceiling operator $\lceil \rceil$ is used in the DDR-ECS formulas for C_s, CI_x, and CF_y.

The optimum length α_{opt} of inter-symbol separator is reduced from 4 for ECS to 2 for DDR-ECS, as shown in Table 4.1(b). The α_{opt} parameter ensures both successful packet reception and maximum data rate. Below this optimal value, the receiver would fail to decode the packet successfully because of the ambiguity between edge spacing and inter-symbol separations. A value of α larger than

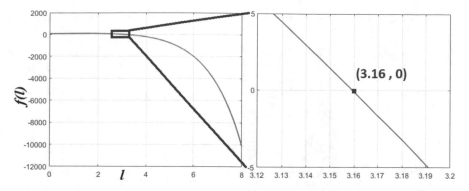

Fig. 4.5 $f(l)$ vs. l

$\alpha_{opt} = 2$ will increase the tolerance to clock variations but decrease data rate. To find the optimum segment length that maximizes data rate, we minimize the number of clock cycles needed to transmit the DDR-ECS packet. Assuming that the data and flag bits are equally likely to be 0 or 1, the expected values of CI_x, CF_y, and C_s are given in Table 4.1(b). Using these expected values and $N = B/l$ in equation for C_t, and defining the function $f(l) \overset{\Delta}{=} \partial E[C_t]/\partial l$, we get the $f(l)$ expression shown in Table 4.1(b). The optimal value l_{opt} can be determined by finding the roots of the non-linear equation $f(l) = 0$. A graphical method to find the optimal segment length l_{opt} for a given α is to plot $f(l)$ and find the l-axis intercept point. Such a plot is shown in Fig. 4.5 for $\alpha = 2$, which results in $l_{opt} = 3.16 \approx 4 \quad bits$. The value of l_{opt} is the same as for ECS, which means that using both edges in edge coding does not affect the inherent properties of edge-coding signaling techniques. Segments smaller than 4 bits reduce data rate due to the increased number of inter-symbol separators. On the other hand, segment lengths larger than 4 bits affect data rate negatively due to the increase in MSB index numbers. To achieve maximum data rate, a segment length of 4 bits has been selected.

4.5 Experimental Verification and Results

The DDR-ECS transceiver hardware discussed in Sect. 4.3 is realized using Verilog HDL. A full experimental setup comprised of two nodes is implemented over a Xilinx FPGA platform. Each node includes a DDR-ECS transmitter, a receiver, and control logic. A clock speed of 25 MHz is used at both ends of the communication link. The selection of 25 MHz is just for illustration purposes, and much higher clock rates can be used. It is to be noted that the ECS data rate increases linearly with the clock rate. The control logic at Node1 sends the 16-bit data starting at 0 with an increment of 1 at each transmission. To verify the complete round-trip chain, the

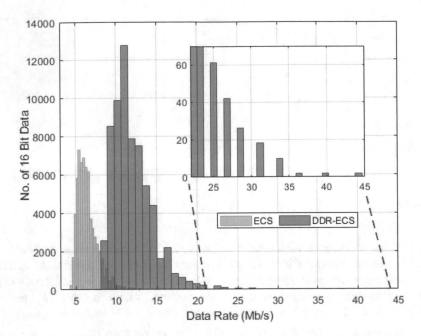

Fig. 4.6 Data rate histograms at 25 MHz clock

control logic at Node2 sends back the received data to Node1 where the original and returned data words are compared. The implemented DDR-ECS transceiver is also synthesized using GLOBALFOUNDRIES 65 nm CMOS technology. The synthesis results are shown in Table 4.1(c). It is clear from the results that double-edge-coded signaling has approximately doubled the data rate without impacting the inherent properties and the performance of the ECS protocol. DDR-ECS offers dynamic data rates in the range of 7.8–44.4 Mb/s (averaging 12 Mb/s) with a 25 MHz of a clock. The data rate histogram for a 16-bit DDR-ECS is shown in Fig. 4.6. The power consumption is unchanged, but the energy-per-bit is improved approximately by a factor of two. Interestingly, the gate count for the DDR-ECS transceiver is slightly less than the ECS one. As compared to NRZ serial transfer using the state-of-the-art low-power CDR, DDR-ECS reduces power consumption by more than 79% and area by more than 88% as shown in Table 4.2.

Table 4.2 DDR-ECS comparison with simple serial

	Power (μW)			Area (gate count)			
	SRL[a]	CDR	Total[d] (PI)[e]	SRL	CDR[c]	Total[d] (PI)[e]	
DDR-ECS	19	N/A	19	1943	N/A	1943	65 nm
NST[b]	32.1	70	102.1 (81%)	1327	15,600	16,927 (88%)	90 nm [37]
		62.5	94.6 (80%)		60,000	61,327 (97%)	90 nm [38]
		90	122.1 (84%)		N/A	N/A	90 nm [12]
		57.5	89.6 (79%)		19,800	21,127 (91%)	65 nm [77]
		60.6	92.7 (80%)		N/A	N/A	28 nm [73]

[a] Serializer
[b] NRZ serial transfer
[c] Estimated calculation
[d] SRL+CDR
[e] %Increase as compared to DDR-ECS

4.6   Conclusions

As compared with the standard single-edge-coded signaling, the proposed double data rate, edge-coded signaling, DDR-ECS, doubles the data rate without any noticeable impact on the ECS power and area budgets. The proposed hardware implementation of the DDR-ECS transceiver was shown to preserve the same power consumption and a similar form factor as ECS. Additionally, the built-in features of secure and reliable ECS communication are preserved.

Chapter 5
Power Management

There is no energy crisis, only a crisis of ignorance.

R. Buckminster Fuller

Chapter 2 has reported on the power consumption of the three generations of ECS based on their *functional logic* only but ignored the power consumed by the *physical layer* (PHY). This layer is comprised of two tri-state buffers and a pull-down resistor connected to the single wire. Although the ECS family itself is power-efficient, the resistance can be the source of significant power consumption. The goal of this chapter is twofold. First, we address the gap in the ECS family power analysis by including the PHY layer in the overall power consumption. Second, we show how to reduce the power consumed in the PHY layer by controlling the width of the transmitted pulses. Furthermore, a mathematical model is developed and used to derive rigorously the performance parameters related to ECS PHY power management. The mathematical model is also used to derive a lower bound on the width of the pulses and therefore a lower bound on the power consumed in ECS PHY. The new power control policy is applied to a single-wire link with significant power saving achieved above and beyond the clock and data recovery. These power savings are obtained without any impact on data rate and bit error rate (BER). The implementation of power management technique uses ECS1, while the methods and results are valid to the whole ECS family.

5.1 ECS1 Power Management

5.1.1 Sources of Power Consumption

The ECS1 power consumption reported in Chap. 2 addresses only the ECS1 functional logic but ignores the power consumption in the PHY layer. For instance, the resistor connected to the single wire may be an important source of power

© Springer Nature Switzerland AG 2022
S. Muzaffar, I. M. Elfadel, *Secure, Low-Power IoT Communication Using Edge-Coded Signaling*, https://doi.org/10.1007/978-3-030-95914-2_5

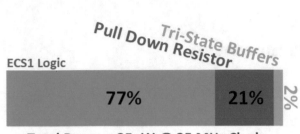

Fig. 5.1 ECS1 transceiver power consumption (50% duty cycle)

consumption in a full transceiver system. Taking ECS1 as an example in Fig. 5.1, the resistor contributes almost $\sim$7 μW to the total ECS1 power dissipation (21% with 50% duty cycle of pulses). This is rather significant, and opportunities for reducing the resistor power dissipation must be explored. This is the main objective of the following subsections.

5.1.2   Proposed ECS1 PHY

Approach

ECS1 is based on transmitting pulses on a single wire. The pull-down resistor attached to the single wire has power dissipation that is linear in the pulse duty cycle, λ. One approach to reduce PHY power is therefore to reduce the pulse duty cycle. This can be done within the pulse generation circuit with the constraint that the overhead in pulse width control should be much less than the projected power savings resulting from the narrower pulse. One approach to pulse generation is to AND the input pulse of duty cycle λ_{In} with its delayed version λ_D, where the delayed pulses are obtained using buffers as illustrated in Fig. 5.2a. More buffers result in a smaller λ at the output of the AND gate. The power overhead of such approach will of course increase with the number of inserted buffers and will therefore negate any power saving we may expect from a smaller λ. Another possible approach is to use a delay line [30, 45] instead of the inserted buffers, but again the power consumption of such delay line outstrips the $\sim$7 μW power budget and so it is unable to achieve the narrow λ that is needed to reduce PHY power consumption.

To meet the power consumption requirements, an inverter coupled with a small capacitor is used to generate narrow pulses of λ duty cycle. This pulse generation and control circuit is shown in Fig. 5.2b. The single inverter is used to invert and delay the input pulse λ_{In} with the capacitor providing an additional load to increase the total delay. When λ_{In} and the delayed inverted pulse λ_{ID} pass through the

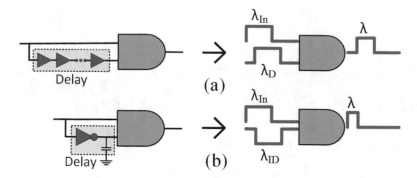

Fig. 5.2 Duty cycle generation. (**a**) Buffers approach. (**b**) Inverter and capacitor approach

Fig. 5.3 Proposed ECS1
PHY layer

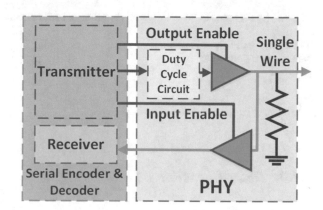

AND gate, a narrow λ is generated. As shown in Fig. 5.3, the proposed ECS1 PHY includes the duty cycle generation circuit at the back of the tri-state buffer.

PHY Circuit Implementation

The proposed ECS1 PHY circuitry is implemented using a 45 nm process. The circuit diagram is shown in Fig. 5.4. The delay inverter is comprised of the $M1$ and $M2$ transistors loaded with a delay capacitor C_D. The AND operation is performed using transistors $M3 - M8$, and the output is coupled with the pull-down resistor R_{PD} through a tri-state buffer. Table 5.1 catalogs the used parameter values. In the following subsections, we describe how to select the values of capacitor C_D and resistor R_{PD} to achieve a given duty cycle.

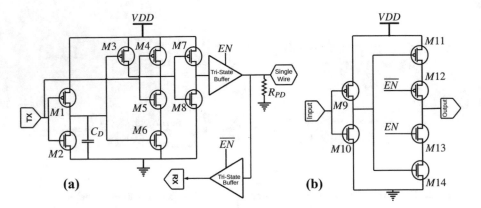

Fig. 5.4 (**a**) Proposed ECS1 PHY circuit (**b**) Tri-state buffer circuit

Table 5.1 Design
parameters of duty cycle
generator

Parameters	Proposed	Unit
NMOS (W/L)	120/45	nm/nm
PMOS (W/L)	240/45	nm/nm
C_D (min)	4.6	fF
R_{PD} (min)	53.28	KΩ
λ_{In}	20	ns

5.1.3 Delay Capacitor

The pulse duty cycle λ is the time difference between the rising edge of the input
pulse λ_{In} and the falling edge of the inverted and delayed pulse λ_{ID}. The time
constant τ for the inverted output signal of the delay block (shown in Fig. 5.2b) is
determined by the delay capacitor C_D and the ON resistance R_{on} of the transistor,
$\tau = C_D R_{on}$. The smaller the time constant τ, the faster the discharging of C_D, and
the narrower the λ. However, C_D cannot be reduced to zero as pulse detectability
imposes both lower and upper bounds on C_D. The expressions of these bounds will
be given in the next two paragraphs.

Lower Bound on Delay Capacitance

The lower bound on the selection of C_D is imposed by the time constant for the
rising edge of the pulse at the output of inverter block of $M1$ and $M2$. During the
rising edge, the $M1$ PMOS transistor is ON, while the $M2$ NMOS is OFF. The
time constant is then defined by the ON resistance of the PMOS transistor, $\tau =
C_D R_{P-on}$. For a successful pulse detection at the receiver, the pulse voltage level

should be $V_C \geq \frac{2}{3}V_{dd}$, where V_C is the voltage output across C_D:

$$V_C = V_{dd}\left(1 - e^{-\frac{t}{\tau}}\right) \tag{5.1}$$

Using $V_C = \frac{2}{3}V_{dd}$, $t = t_r$, $R_{on} = R_{P-on}$ and solving for C_D, one gets

$$C_D \geq -\frac{t_r}{\ln\left(\frac{1}{3}\right)R_{P-on}} \implies C_D \geq 0.91\frac{t_r}{R_{P-on}} \tag{5.2}$$

where t_r is the rise time during which the capacitor is charged up to $\frac{2}{3}V_{dd}$. In 45 nm CMOS technology, $R_{P-on} = 13.32\,\text{K}\Omega$, which leads to the smallest possible value of $C_D = 4.6\,\text{fF}$. Violating this condition will result in a pulse detection failure at the receiver, that is, $\lambda = 0$ or 0% duty cycle.

Upper Bound on Delay Capacitance

The upper bound on C_D is imposed by the time constant for the pulse falling edge at the output of the inverter block of $M1$ and $M2$. During the falling edge, the $M1$ PMOS transistor is OFF, while $M2$ NMOS is ON. The time constant is then defined by the ON resistance of the NMOS transistor, $\tau = C_D R_{N-on}$. The higher τ is, the larger λ will be. Beyond a certain C_D value, λ saturates at 50%. For a duty cycle less than 50%, the λ_{ID} voltage level should be $\leq \frac{1}{3}V_{dd}$ and should be achieved, while the input pulse λ_{In} is high. In other words, $t_f \leq \lambda_{In}$, where t_f is the output fall time during which the capacitor is discharged down $\frac{1}{3}V_{dd}$ and is determined by the capacitor discharge equation

$$\frac{1}{3}V_{dd} = V_{dd}\,e^{-\frac{t_f}{\tau}} \implies t_f = -\ln\left(\frac{1}{3}\right)R_{N-on}C_D \leq \lambda_{In} \tag{5.3}$$

which leads to the upper bound

$$C_D \leq 0.91\frac{\lambda_{In}}{R_{N-on}} \tag{5.4}$$

In 45 nm CMOS technology, $R_{N-on} = 17.22\,\text{K}\Omega$, which leads to the largest possible value of $C_D = 1\,\text{pF}$ in the case of a 20 ns long input pulse. Beyond this upper bound on capacitance, λ saturates to 50%.

5.1.4   Sizing the Pull-Down Resistor

The pull-down resistor, R_{PD}, attached to the single wire at the output of the tri-state buffer keeps the line low during the high-impedance state of the buffer. R_{PD} directly impacts the output voltage at the output rising edge during which the transistors $M11$, $M12$, and $M13$ are ON, while $M14$ is OFF. A voltage divider comprised of two PMOS ON resistances (i.e., $2R_{P-on}$) and R_{PD} is formed. For the successful detection of the pulse at the receiver, the output pulse voltage V_{Pulse} should satisfy $V_{Pulse} \geq \frac{2}{3}V_{dd}$ where

$$V_{Pulse} = V_{dd}\frac{R_{PD}}{R_{PD} + 2R_{P-on}} \tag{5.5}$$

which means $R_{PD} \geq 4R_{P-on}$. In 45 nm CMOS technology, the $R_{P-on} = 13.32$ KΩ, which leads to the smallest possible value of $R_{PD} = 53.28$ KΩ. Violating this condition will result in small pulse voltages at the output that may fail detection at the receiver. The larger R_{PD} is, the larger the output pulse swing.

5.1.5   Duty Cycle

The pulse duty cycle λ at the output is given by

$$\lambda = t_r + t_f \tag{5.6}$$

which yields

$$\lambda = 1.0986\,C_D\,(R_{P-on} + R_{N-on}) \tag{5.7}$$

The percentage duty cycle, $\lambda_\%$, is determined as

$$\lambda_\% = \frac{\lambda}{2\lambda_{In}} \times 100 \tag{5.8}$$

The minimum duty cycle λ_{min} is determined using the smallest possible delay capacitance (i.e., $C_D = C_{D-min} = 4.6$ fF) in (5.7) and is given as

$$\lambda_{min} = 1.0986\,C_{D-min}\,(R_{P-on} + R_{N-on}) = 154.34\,\text{ps}$$

where R_{P-on} and R_{N-on} are 13.32 KΩ and 17.22 KΩ, respectively, for a 45 nm process. Also,

$$\lambda_{min-\%} = \frac{\lambda_{min}}{2\lambda_{In}} \times 100 = 0.39\,\%$$

5.2 Results

5.2.1 Power Analysis

The proposed policy to manage power consumption, through controlling the pulses duty cycle, is verified rigorously on an ECS1 protocol using Spice-level simulations for a 45 nm CMOS process in the Cadence design environment. The power analysis is performed using the minimum duty cycle of 0.39%, which is generated with $C_D = 4.6\,\text{fF}$, $R_{PD} = 54\,\text{K}\Omega$, and $\lambda_{In} = 20\,\text{ns}$ (i.e., clock = 25 MHz). The resistor, with the proposed PHY, consumes only 0.1% (27.22 nW) of the total PHY power (265.4 nW). The tri-state buffers' power consumption is also reduced to 0.9% (238 nW) only. The total power saving, as compared to the one shown in Fig. 5.1, is 20%. The improved power consumption ratios are shown in Fig. 5.5. The total PHY power consumption is linearly related to the duty cycle as shown in Fig. 5.6a. Clearly, the pulse duty cycle greatly impacts the overall power consumption. For our experimental prototype, the duty cycle generator should not consume more than $7\,\mu\text{W}$ ($\lambda \leq 5\%$ for the experiment performed). On the other hand, the power consumption reduces with the increase in pull-down resistance, as shown in Fig. 5.6b. If the PHY is intended to connect to downstream loads presenting an input capacitance in parallel with the pull-down resistance, then increasing the resistance will increase the rise and fall slews of the pulse. Therefore, the upper bound on the

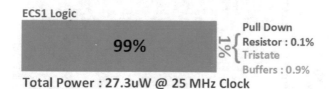

Fig. 5.5 Proposed ECS1 transceiver power consumption (0.4% duty cycle)

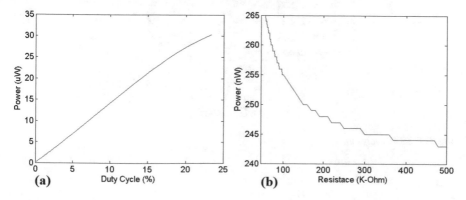

Fig. 5.6 PHY power consumption (**a**) vs. Duty cycle ($R_{PD} = 54$ KΩ) (**b**) vs. Resistance ($C_D = 4.6$ fF)

resistance is application-dependent and should be determined based on the overall timing performance of the full link.

5.2.2   BER Analysis

Noise associated with the off-chip environment can have an effect on data pulses. The increase or decrease in pulse levels, due to the external noise, makes it difficult for the receiver to detect pulses correctly. Depending on the noise level, an extra pulse may be detected or a pulse may be skipped. In both cases, one gets a decoding error. To analyze the performance of ECS1 in the presence of noise, the encoded pulse stream of data is exposed to white Gaussian noise. The noisy signal is filtered at the receiver end, then decoded according to the ECS1 protocol, and the number of errors encountered is counted. The results are plotted in Fig. 5.7 for different values of E_b/N_0 (the ratio of energy-per-bit to noise power spectral density) for both BPSK and ECS1. For ECS1, the results are plotted with variations in the pulse duty cycle. ECS1 is less immune to noise as compared to BPSK, but its BER reduces rapidly to zero at E_b/N_0 of $\sim$10.5 dB. Variations in pulse duty cycle have no significant impact on ECS1 BER.

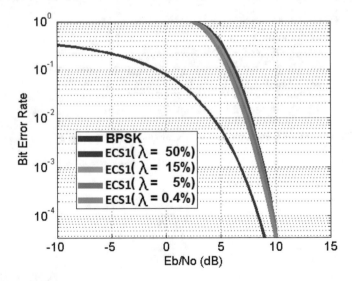

Fig. 5.7   BER analysis

5.3   Conclusions

In this chapter, we have presented a detailed power management policy for ECS protocols to improve further their ultra-low-power characteristics without impacting their data or bit error rates. In this policy, pulse duty cycle control is used to reduce the power consumption of the ECS1 PHY layer. In addition, a mathematical model has been used to select the physical design parameters related to ECS1 PHY power management. Based on our implementation and analysis, a minimum capacitance of 4.6 fF and a minimum resistance of 54 KΩ are recommended to generate the smallest possible duty cycle of 0.39%. These recommended design parameters have been experimentally verified using Spice-level simulations of a 45 nm process. On an ECS1 platform, an additional power saving of 22% has been achieved in the its PHY layer that further improves on the already significant power savings in the ECS1 functional logic as reported in Chap. 2.

Chapter 6
Automatic Protocol Configuration

It would not be long ere the whole surface of this country would
be channelled for those nerves which are to diffuse, with the
speed of thought, a knowledge of all that is occurring
throughout the land, making, in fact, one neighborhood of the
whole country.

Samuel Morse

The objective of this chapter is to provide an algorithm for automatically configuring the ECS protocol parameters at the power-on phase. To achieve this, we introduce a new algorithm that rigorously specifies the protocol configuration procedure and uses closed-form formulas to assign suitable protocol parameters to both ends of the transmission link based on clock-rate differences. Pulse count differences between the pulse trains at the transmitter and receiver are used to find a suitable inter-symbol separator coefficient so as to eliminate the need to know the exact clock rates at both ends of the link. Moreover, a power-efficient realization with very low hardware complexity of the inter-symbol separator coefficient calculations is proposed to enable efficient protocol configuration. The hardware realization is evaluated in both the FPGA and ASIC design flows. This realization of the auto configuration algorithm is based on ECS1. However, the methods and results are valid across the entire ECS family.

6.1 Automatic Parameter Detection

6.1.1 Algorithm

Consider a single-channel link between two IoT devices that are configured in a master–slave topology, as in Fig. 6.1. Each device uses its own clock frequency. Also each device generates a reference clock of 100 KHz along with the local device clock. The reference clock frequency is defined as the default clock for the ECS1 transceivers and can be used anytime to establish a communication link between the

© Springer Nature Switzerland AG 2022
S. Muzaffar, I. M. Elfadel, *Secure, Low-Power IoT Communication Using Edge-Coded Signaling*, https://doi.org/10.1007/978-3-030-95914-2_6

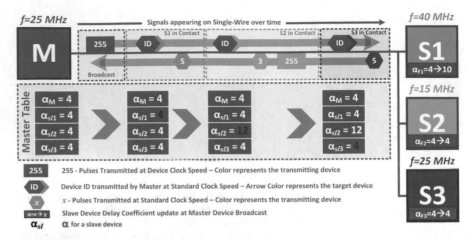

Fig. 6.1 An example of automatic parameter detection

master and slave devices. All the devices power up with the default inter-symbol separator coefficient $\alpha = 4$. The master device also maintains a table of α settings for all the slave devices, and by default all the α entries are equal to 4. The master device selects a corresponding α for a slave device from the table and considers this as its own α when it needs to communicate with that particular slave. The objective of the algorithm is to update not only the master table entries but also the local α settings of all the connected slave devices. These updated values are then used to carry out a successful communication at the local clock rate considered as a ECS1 clock at each end of the link. The ECS1 α value for the reference clock is $\alpha = 4$ irrespective of the local settings in the slave device and its table entries in the master device.

In the power-on phase, the master device transmits system clock pulses for one second to all the slave devices. At the first rising edge of the transmitted pulse stream, the slave devices start counting both the received pulses f_S and their own device clock pulses f_F. The counting ends on pulses from the master device. Using the pulse count difference ($\Delta f = f_F - f_S$), the high-speed node of the link at f_F is recognized. The slave sets its local α_F to a new *suitable* value if it is the high-speed node. Otherwise it keeps the default value of 4. In reference to Algorithm 4, these steps are shown on line 2 for the master device, and lines 1 to 14 for the slave device. The process to find a suitable inter-symbol separator coefficient value is presented in the next subsection. Next, the master device starts the second part of parameter detection by asking each slave device about the status of parameter setting. This process is carried out using the reference clock of 100 KHz. The master device sends a slave ID followed by the reception of either 5 or 3 pulses from the target device. If the target slave was successful in setting a new α_F, 5 pulses would be received and the master will update the table entry for this device with $\alpha = 4$. If the target slave has failed in setting a new α_F, 3 pulses at 100 KHz would be received followed by the reception of system clock pulses of the slave device for one second. The 3 or 5 pulses in response to slave ID are followed by a 6 clock cycle separator to distinguish these pulses correctly from the following system clock pulses. In Algorithm 4, these steps are shown on lines 4 through 9 for the master device, and lines 15 through 20 for the slave device. On reception of these pulses, the master device follows the same counting and α_F setting process as described for the slave and updates the table entry for this slave with a new suitable α_F. The calculation steps are shown by lines 10 to 17 for master device, and lines 7 to 13 for slave device in Algorithm 4. The process continues until all the slave devices are covered. The master sets one α entry in its table for each slave in the network, as shown on lines 7, 12, and 17 of Algorithm 4 for the master device. The power-on automatic detection process allows the master device to configure all slave devices before they start communication using their device clocks. An example of such configuration process for a single-channel link with three slave devices is illustrated in Fig. 6.1.

6.1.2 Inter-symbol Separator Coefficient Calculation

Given the frequency information of the slow node, the inter-symbol separator coefficient α is always calculated by the fast node. This is because the fast node can adjust its separator coefficient to reduce its transmission speed, but it is impossible for the slow node to go beyond its device clock rate. To find a suitable α, first we need to set the maximum supported clock ratio β_{max}. The phase shift φ and the clock jitter ψ are the knobs in Eq. (3.8) to adjust the supported percentage of clock ratio β_{max} assuming the default values of $\alpha_F = 4$ and $w_F = 0.5$. In the ideal case where both φ and ψ are zero, the ratio reaches to $\approx 41\%$. However, this is not the case in real world applications where there are always a phase shift and clock jitter that reduce the clock ratio. Empirical results and the information provided in [57] show that the safer values of $\varphi = 1$ and $\psi = 0.01$ lead to a ratio of $\approx 18\%$. Once the frequency information of the slow node is shared with the fast node, the β_{max} is calculated using (3.8). Equations (3.9) and (3.10) are then used to find the suitable α_F as per the steps given in Algorithm 4. The plot in Fig. 6.2 shows the variation in α_F due to the increase in fast-node frequency f_F when the slower node is at $f_S = 25\,\text{MHz}$.

While Algorithm 4 is suitable for finding α_F, it does have an important limitation. Indeed, to share the precise frequency information, a device needs to transmit pulses at least for one second. This information exchange tends to be time consuming when there are many devices attached to a single-channel link. Therefore, instead of transmitting the pulses for one second, the transmission of just 255 pulses is adopted from one node to another. Hence, at the receiver node, both the received pulses count, N_S, and the device clock pulses count, N_F, become known. Algorithm 4 applies without any change in the formulation except replacing the f_F, f_S, and f_{FN} with N_F, N_S, and N_{FN}, respectively. This method of calculating α_F by considering solely the pulse counts is efficient and suitable for all IoT devices including low-end microcontrollers.

Algorithm 4 Automatic parameter detection

Legends:
- f_S : Received Pulse Count[1]
- f_F : Device's Local Clock Pulse Count[1]
- f_{100} : 100 KHz reference Clock

─────────────────────────── **Master Device Algorithm** ───────────────────────────

Inputs:
- $f_{d,master}$: Master Device Clock
Outputs:
- $\alpha_{sl}[n]$: α settings of n Slave Devices in Master Device Table

---

1: **Set** $n = \# \, of \, slaves, i = 0$
2: **Broadcast** pulses for one second[2] at $f_{d,master}$
3: **while loop**$(i < n)$
4:       **Send** $Slave\_ID[i]$ at f_{100}
5:        $Pulses = $ **Receive** Pulses at f_{100}
6:       **if** $(Pulses = 5)$
7:            $\alpha_{sl}[i] = 4$
8:       **else**
9:            $f_S = $ Receive Pulses at $f_{d,slave}$ for one second[2]
10:            $f_F = f_{d,master}$
11:           **if** $(f_F \le f_S \beta_{max})$
12:                $\alpha_{sl}[i] = 4$
13:           **else**
14:                $f_{FN} = f_F - (\beta_{max} - 1) \, f_S$
15:                $\beta = f_{FN}/f_S$
16:                $A_F = \lceil 4 \, w_F \beta^2 + \beta \, (\varphi + \psi) \rceil$
17:                $\alpha_{sl}[i] = mod(A_F, 2) + A_F$
18:        $i++$
19: **end loop**

─────────────────────────── **Slave Device Algorithm** ───────────────────────────

Inputs:
- $f_{d,slave}$: Slave Device Clock
Outputs:
- $\alpha_{F,n}$: Local α Setting of nth Slave Device

1: $f_S = $ **Receive** pulses for one second[2] at $f_{d-master}$
2: $f_F = f_{d,slave}$
3: **if** $(f_F < f_S)$
4:       $\alpha_{F,n} = 4$
5:       **Setting** Failed
6: **else**
7:       **if** $(f_F \le f_S \beta_{max})$
8:            $\alpha_{F,n} = 4$
9:       **else**
10:            $f_{FN} = f_F - (\beta_{max} - 1) \, f_S$
11:            $\beta = f_{FN}/f_S$
12:            $A_F = \lceil 4 \, w_F \beta^2 + \beta \, (\varphi + \psi) \rceil$
13:            $\alpha_{F,n} = mod(A_F, 2) + A_F$

(continued)

Algorithm 4 Automatic parameter detection

14:     **Setting** Passed
15: **Wait** for Slave ID at f_{100}
16: **if** (Setting Failed)
17:     **Send** 3 Pulses at f_{100}
18:     **Send** pulses for one second[2] at $f_{d,slave}$
19: **else**
20:     **Send** 5 Pulses at f_{100}

[1] See Sect. 6.1.2 for more specific settings.
[2] A more efficient alternative is proposed in Sect. 6.1.2.

Fig. 6.2 α_F vs. f_F when $f_S = 25\,\text{MHz}$

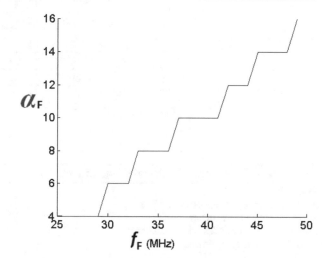

6.1.3   Low-overhead Hardware for α_F Calculation

The calculations to find a suitable α_F can be readily implemented in a micro-controller using the C language. On the other hand, for a VLSI or an FPGA implementation, the floating point operations involved in such calculations present a significant area and power overhead due to the additional hardware resources needed to handle floating point arithmetic. To mitigate this overhead, we propose a hardware decoding scheme that maps the pulse count difference ($\Delta N = N_F - N_S$) to a suitable α_F.

The HW decoder works as follows. For each α_F, there exists a range of ΔN as shown in Table 6.1. Lines 3 to 9 in Algorithm 5 illustrate the process of finding these ranges. Note that in Line 1 of Algorithm 5 N_S is set to 255 because α_F is always calculated by the fast node, and it is guaranteed that 255 pulses will be transmitted by the slow node. Figure 6.3 shows the numerical result, with $f_S = 25\,\text{MHz}$ and $f_F \in [25\,\text{MHz}, 125\,\text{MHz}]$, generated through looping statements in Algorithm 5, Lines 2, and 10 to 12. The number of ranges of α_F sets an upper bound on the maximum clock ratio supported by the automatic baud rate feature. The sweep ranges for f_F can be increased or decreased as per the application requirement. A custom HW decoder is designed that accepts a 10-bit $\triangle N$, finds the corresponding range as per Table 6.1, and sets an 8-bit α_F at the output.

Table 6.1 α_F ranges for decoder implementation

$\triangle N_{min}$	$\triangle N_{max}$	β Clock ratio (%)	α_F
1	47	≤ 18.4	4
48	74	≤ 28.8	6
75	122	≤ 47.6	8
123	164	≤ 64.0	10
165	202	≤ 79.2	12
203	238	≤ 93.2	14
$\vdots$	$\vdots$	$\vdots$	$\vdots$
997	1009	≤ 395.6	96
1010	1020	≤ 400.0	98
$\vdots$	$\vdots$	$\vdots$	$\vdots$

Algorithm 5 α_F ranges for decoder implementation

Inputs:
- β_{max} : Maximum supported clock ratio
- f_S : Slow node frequency
- f_{Fmax} : Maximum fast-node frequency
- $\triangle f$: Frequency step value

Outputs:
- $\triangle N_{min}$: Minimum device clock pulse count
- $\triangle N_{max}$: Maximum device clock pulse count

1: **Set** $N_S = 255$, $T_S = (1/f_S) \times N_S$, $f_{F=f_S}$, and $i = 0$
2: **while loop** ($f_F \leq f_{Fmax}$)
3:      $N_F = T_S \times f_F$
4:      **if** ($N_F \leq N_S \beta_{max}$)
5:          $\alpha_F [i] = 4$
6:      **else**
7:          Find N_{FN}, β, A_F
8:          $\alpha_F [i] = mod(A_F, 2) + A_F$
9:      $\triangle N [i] = N_F - N_S$
10:     i++
11:     $f_F = f_F + \triangle f$
12: **end loop**
13: **Find** $\triangle N_{min}$ and $\triangle N_{max}$ from $\{\triangle N\}$ for each of the unique α_F values in $\{\alpha_F\}$

Fig. 6.3 α_F vs. $\triangle N$ when $f_S = 25\,\text{MHz}$ and $f_F \in [25\,\text{MHz}, 125\,\text{MHz}]$

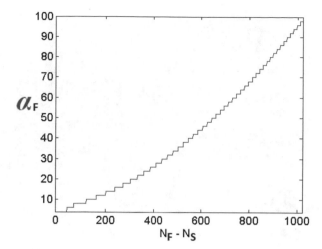

6.2 Experimental Verification

An experimental setup comprised of two IoT nodes communicating using the ECS1 protocol is used to verify the operation of the automatic parameter detection algorithm. Each node is composed of a ECS1 protocol module, an Automatic Parameter Detector (APD), a ECS1 PHY layer, PHY switch, and a clock generator, as shown in Fig. 6.4. The clock generator provides two clocks to a node, a device clock and a reference clock (100 KHz). The PHY switch, controlled by APD, is used to allow either the ECS1 module or the APD to access the PHY layer for establishing the physical link. Both the master and slave nodes have a similar implementation except for the differences present in APD module. The master APD is composed of three main modules: a controller, a decoder, and an α_F table. The slave APD is composed of the same modules except that there is only one local α_F setting instead of a table. The controller module takes care of all the power-on configuration process, controls the PHY access, and connects the ECS1 to a suitable clock during communication. The decoder accepts $\triangle N$ as an input that is generated by the controller and outputs a suitable α_F for table entry at the master node or for local setting at the slave node. During communication after completing the power-on configuration process, the master node directs a particular slave device using the 100 KHz reference clock, fetches the corresponding α_F from the table, and using this α_F communicates further with the said slave device. This on-the-fly change in the inter-symbol separator coefficient allows the slave devices to use their device clock to adjust their data rates and thus, an adaptive baud rate network is established.

The full experimental setup is implemented in Verilog on the Xilinx Virtex-7 FPGA platform. Two separate clocks with different frequencies and phase shifts, one for each node, are generated with the help of a Virtex-7 on-chip PLL. The master clock is fixed at 25 MHz while the clock rate for the slave node is allowed to deviate from 25 MHz. The master does power-on configuration and then verifies two-way communication using the automatically detected parameters.

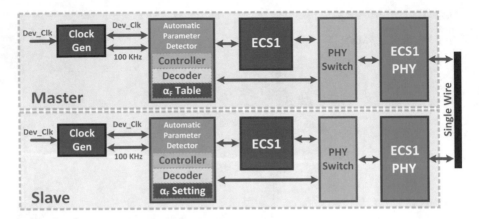

Fig. 6.4 Block diagram of the experimental setup

Table 6.2 Synthesis results

	Power (μW)			Area (*gate count*)		
APD system (excluding ECS1)	4.35			≈1500		
ECS family	ECS1	ECS2	ECS3	ECS1	ECS2	ECS3
	26.6	25	19	2356	2150	2098
Total	30.95	29.35	23.35	≈3856	≈3650	≈3598

The experiments confirm that the ECS1 transmission works flawlessly. Along with the FPGA prototype, we have also synthesized the automatic parameter detection system using a Synopsis logic-synthesis flow and a GLOBALFOUNDRIES 65 nm process. We found out that the system shown in Fig. 6.4 (excluding ECS1) maintains the low-power operation of ECS1 consuming only 4.35 μW with a gate count of ≈1500 at a clock frequency of 25 MHz. The synthesis results are shown in Table 6.2.

6.3   Conclusions

The power-on algorithm for automatically detecting the ECS1 protocol parameters allows the master device to configure all the slave devices connected to a single-channel network prior to the start of any device-to-device communication. The proposed automatic configuration methodology eliminates the need for compile-time or manual assignment of protocol parameters. Moreover, the methodology removes the restriction on all the devices in the network to agree on a single communication speed and allows the devices with different capabilities to communicate reliably. Toward this end, the master device selects, in real time, the right ECS1 parameter values that enable it to communicate with a particular slave. This makes the ECS1 single-channel network behave as an adaptive baud rate network. The proposed architecture is experimentally verified and tested on a point-to-point communication link using a Xilinx Virtex7 FPGA platform that illustrates the simplicity, efficiency, and reliability of using automatic ECS1 parameter detection. In particular, we show that the efficient hardware realization of the algorithm maintains the low-power operation of ECS1 protocol while consuming only $4.35\,\mu W$ (65 nm process). Our work has assumed that all slave devices have already been discovered and identified by the master device. The issue of device discovery in ECS1 networks is still open and very much worth investigating.

Chapter 7
Secure ECS Communication

> *Encryption works. Properly implemented strong crypto systems are one of the few things that you can rely on.*
>
> Edward Snowden

This chapter highlights the advantages of a tight integration between the IoT communication protocol on the one hand and lightweight cryptography on the other. This is illustrated in the multilayer integration of edge-coded signaling (ECS) with a novel, parallel, low-latency version of the A5/1 keystream cipher. This integration has resulted in a secure communication protocol that is very well adapted to constrained IoT nodes. The secure ECS solution features both confusion and diffusion defenses while providing both data confidentiality and packet authentication. The secure ECS system has been prototyped on embedded microcontroller, FPGA, and ASIC platforms with all the prototypes confirming the low overhead of the low-latency A5/1 crypto block. The secure ECS prototypes have achieved data rate, power consumption, small footprint figures that are all in line with the original ECS attributes.

7.1 Introduction

The Internet of Things (IoT) offers advanced machine-to-machine connectivity among edge nodes where low-end, low-cost devices are ubiquitous and would constitute, given their crypto vulnerability, a major challenge for a fully secure network. There are many protocols for IoT machine-to-machine communication [2, 64] with the standard practice being to secure the transmission using data block encryption. The encrypted data is then transmitted through the channel-specific transceivers and is decrypted at the receiver end. In some cases, nested encryption is applied to strengthen data security. As shown in Fig. 7.1, regardless of the communication links the modulation scheme or the encryption technique used in the transceivers, the physical signals always represent the packet bits, either 0 or

© Springer Nature Switzerland AG 2022
S. Muzaffar, I. M. Elfadel, *Secure, Low-Power IoT Communication Using Edge-Coded Signaling*, https://doi.org/10.1007/978-3-030-95914-2_7

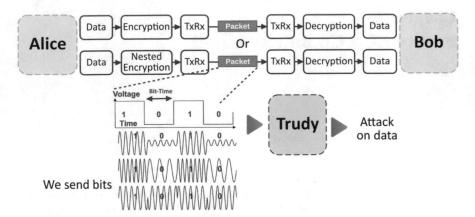

Fig. 7.1 Block diagram of a standard digital communication link between Alice and Bob under a cryptographic attack by Trudy

1. Trudy, the intruder, can listen in and receive all the packets and apply attacks on data directly. In this chapter, we revisit this standard architecture and present a novel architecture that prevents Trudy from even receiving a valid packet in the first place. This has been achieved by establishing a synergy between the communication protocol and the crypto algorithm, where they are enabled to strengthen each other to achieve higher levels of secure communication for constrained IoT nodes. The new crypto design is based on three innovations. The first is a new data signaling technique, called Edge-Coded Signaling (ECS), that is dynamic yet ultra-low power. The second is an accelerated lightweight crypto algorithm with a one clock cycle overhead and a tiny hardware footprint. The third innovation is a multilayer security protocol in which both data and signaling information are secured with the help of the communication technique's dynamic features.

Given the dynamic nature of ECS, the transmitted ECS packet contains no raw bits but enough information about the *support* of the data block so as to enable its reconstruction at the receiver. At the physical level, the ECS packet is transmitted as a sequence of pulse streams where each pulse stream represents an aspect of the data block support. These aspects include (1) a *reception guide* which acts as a table of content that facilitates the reception of the subsequent symbols; (2) an *assembly guide* which enables the assembly of received symbols in their proper order, and (3) a *transformation guide* which helps reverse any possible transformations the original data block may have been subject to (e.g., one's complement). The ECS packet content is dynamic, and even a single-bit change in the original data block may drastically modify the ECS packet. When properly used, such property in the *signaling* protocol should enhance the *diffusion* property of the cryptographic algorithm. This is one illustration of what we meant by the synergy between signaling and encryption. The reader is referred to Chap. 2 for more information on the ECS signaling protocols.

Now there has been a significant amount of work in the field of lightweight IoT cryptography [31, 47]. Since the adoption of the Advanced Encryption Standard (AES) in 2002, many lightweight symmetric key *block* ciphers have been developed. They include XTEA, CLEFIA, FeW, HIGHT, LBlock, PRESENT, and Piccolo [8, 24, 34, 71, 72, 74, 80]. The focus of these developments has been on improving the crypto algorithm itself independent of other considerations as may arise in a full IoT communication link. The paramount emphasis of the IoT cryptographic research has of course been for the lightweight crypto to ensure reasonable defense against malicious attacks on constrained devices. This chapter advocates that such defense can be significantly improved if the dynamic and diffusivity properties of the ECS signaling protocol are seamlessly incorporated into the cryptographic algorithm. Additionally, this chapter highlights that even "weak" lightweight cryptographic algorithms can become quite competitive in terms of their cryptanalysis when used in conjunction with ECS. To illustrate such an approach to improving lightweight crypto security, we have selected the venerable GSM cipher, A5/1 [74], and shown how its attack complexity can be increased by a 2^{20} factor while reducing its latency overhead to one clock cycle. We call improved A5/1 cipher HSA5/1 (High-Speed A5/1).

Another important aspect of IoT cryptography is packet authentication. This is achieved in our ECS-based implementation using packet identification signatures that get attached to the data support block with the two encrypted together using HSA5/1. Finally, as mentioned earlier, the ECS packet contains information on how to reconstruct the bit stream from its support. Such information is also encrypted using HSA5/1 and a key different from those used to generate the identification signatures or encrypt the data support block. This provides an additional obfuscation layer of ECS transmission security and helps protect the ECS packet at Trudy's end. In reference to the block diagram of Fig. 7.2a, the secure ECS protocol comprises the following steps:

1. *Key Generation*: The architecture uses a strengthened lightweight key generation technique to produce one main keystream and several subkeys. The keys are updated every transaction at both ends of the communication link.
2. *Data Identification*: The architecture embeds one subkey as an identifier within the original data to prepare the ECS dataframe.
3. *Data Encryption*: The architecture encrypts the ECS data using the main keystream.
4. *ECS Encoding*: The architecture encodes the encrypted bitstream using the ECS protocol and, as a byproduct, generates the ECS encoding information pertaining to packet configuration.
5. *Packet Encryption*: The ECS encoding information is then encrypted using the remaining subkeys.
6. *Transmission Obfuscation*: To transmit the encrypted ECS frame (data and configuration), the architecture builds the packet using the encrypted ECS encoding information but runs the transmission process using the non-encrypted version.

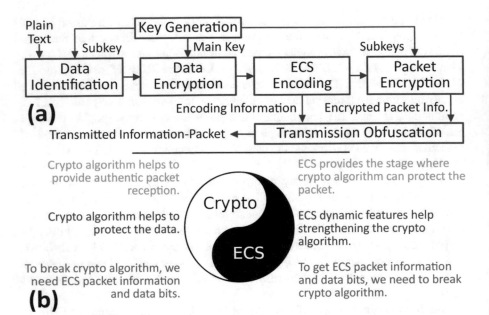

Fig. 7.2 (**a**) Block diagram of secure edge-code-signaling (ECS) protocol showing the six steps of key generation, identification, data encryption, ECS encoding, ECS packet header encryption, and transmission obfuscation. (**b**) The yin and yang of signaling and encryption

Such a lightweight, multilayer cryptographic technique provides a hard-to-break challenge to an attacker while preserving ECS performance. The major features of this combination of an IoT signaling protocol with a lightweight cipher are further summarized in Fig. 7.2b as a yin yang figure. The complementarity between signaling and encryption is most evident in the fact that for an unauthorized receiver to decode the ECS packet, she needs to break the encryption, but to break the encryption, the data-dependent ECS encoding parameters need to be known. The secure ECS can be further protected by increasing the number of keys and the number of bits in each key with minimal impact on ECS communication performance. The latter is because of the novel parallel architecture that has been introduced for the crypto algorithm. In essence, the interplay between ECS and the encryption results in making the ECS transmission more secure than if the encryption were used alone in a single-layer cryptographic system. The novelties of this technique are summarized as follows:

1. *High-speed A5/1*: A novel high-speed version of the well-known A5/1 cipher is designed, implemented, and integrated with ECS. This novel symmetric keystream cipher is fully evaluated in terms of its performance, power, and footprint impact on ECS.
2. *Multilayer architecture for secure ECS communication*: This architecture fully integrates the high-speed A5/1 cipher within the ECS encoding and transmission

process. It provides multiple hard-to-break challenges to a malicious attacker with several layers of encryption and transmission obfuscation.

3. *Hardware implementations of high-speed A5/1 and secure ECS*: These implementations have enabled the realistic evaluation of the multilayer, secure ECS architecture from the viewpoints of data rate, power, and footprint. They have amply illustrated the advantageous synergy between the ECS communication protocol and the HSA5/1 cryptographic algorithm. To the best of our knowledge, this is the first time such a synergy has been achieved.

4. *Versatile networking options*: Various network topologies can be supported with the secure ECS communication system, including master–slave, star, ring, and peer-to-peer topologies.

7.2 Proposed Multilayer Secure Communication Architecture

7.2.1 Re-architecting A5/1 for ECS

As mentioned earlier, we have selected A5/1 [74] as an example to illustrate how dynamic features of ECS can be used to strengthen an otherwise weak encryption that has been the subject of numerous attacks [5]. Moreover, in later subsections, we will show that the strengthened crypto not only secures packet reception but also enjoys an additional layer of protection that makes it even more resistant to attacks. The textbook version of A5/1 generates one key bit per iteration that is XORed with 1 bit of data. For a 16-bit data block, a straightforward hardware implementation of A5/1 requires at least 16 iterations to encrypt the data in an ECS packet. Assuming the keystream generator is clocked at one clock cycle per bit, the crypto latency is 16 clock cycles, which would negatively impact the ECS data rate. We address the crypto latency issue by proposing a modification of A5/1 that will generate one 16-bit main key and five 4-bit subkeys in an only clock cycle. The modified A5/1 is called high-speed A5/1 or HSA5/1.

Conventional A5/1

Denote by s the one keystream bit that the conventional A5/1 cipher generates at each iteration and uses it to encrypt one plaintext bit, d_i, to produce one encrypted bit of ciphertext $c_i = d_i \oplus s$. To generate the ciphertext, A5/1 makes use of three registers, X, Y, and Z, that are 19-bit, 22-bit, and 23-bit wide, respectively. The registers are initially filled using a 64-bit secret key, K, as shown in Table 7.1. The specific steps involved in generating one keystream bit s are given in Algorithm 6. Assuming a straightforward implementation where one clock cycle is consumed

Table 7.1 Encryption comparison between conventional A5/1 VS. high-speed A5/1

	Conventional A5/1	High-speed A5/1
Properties		
Key (K)	64 bits	148 bits
Reg. X	$x_0, x_1, \ldots x_{18}$	$x_0, x_1, \ldots x_{40}$
Reg. Y	$y_0, y_1, \ldots y_{21}$	$y_0, y_1, \ldots y_{42}$
Reg. Z	$z_0, z_1, \ldots z_{22}$	$z_0, z_1, \ldots z_{43}$
		5 Additional 4-bit Registers
Keystream generation equations		
m	$maj(x_8, y_{10}, z_{10})$	$maj(x_{15}, y_{20}, z_{20})$
t_x	$x_{13} \oplus x_{16} \oplus x_{17} \oplus x_{18}$	$x_{33} \oplus x_{38} \oplus x_{39} \oplus x_{40}$
t_y	$y_{20} \oplus y_{21}$	$y_{41} \oplus y_{42}$
t_z	$z_7 \oplus z_{20} \oplus z_{21} \oplus z_{22}$	$z_{14} \oplus z_{41} \oplus z_{42} \oplus z_{43}$
	$s = x_{18} \oplus y_{21} \oplus z_{22}$	$s_{i_S} = x_{i_X} \oplus y_{i_Y} \oplus x_{i_Z}$

Algorithm 6 One iteration of the conventional A5/1

Inputs:
– X, Y, and Z: A5/1 registers
– d_i: i_{th} Data bit to encrypt
Outputs:
– c_i: i_{th} Encrypted data bit

1: Use majority function $maj(\ldots)$ to find m
2: **if** $m == x_8$ **then**
3:     $X = X >> 1$ // (from LSB to MSB)
4: **end if**
5: **if** $m == y_{10}$ **then**
6:     $Y = Y >> 1$ // (from LSB to MSB)
7: **end if**
8: **if** $m == z_{10}$ **then**
9:     $Z = Z >> 1$ // (from LSB to MSB)
10: **end if**
11: Find t_x, t_y, and t_z
12: $x_0 = t_x, y_0 = t_y, z_0 = t_z$
13: Find s
14: $c_i = d_i \oplus s$

---

per bit of ciphertext, then n clock cycles are needed to encrypt a full data block of n bits. This will of course significantly degrades the ECS transmission data rate and increases the power-on latency. A naive approach to address this issue is to use hardware parallelism by replicating the A5/1 logic block n times in order to achieve one clock-cycle operation per bit. However, this approach would require n different keys as it is not possible to use the same key for all the A5/1 replicas. Furthermore, the n A5/1 replica blocks and a separate register set for each block would substantially increase the ASIC silicon area or FPGA resources. In the following section, we introduce a modification of A5/1 that achieves one clock-cycle operation for n-bit keystream generation without A5/1 block replication. We shall also show how the interplay of ECS encoding and HSA5/1 encryption results in making the ECS transmission much more secure than if HSA5/1 were used alone in a single-layer cryptographic system.

Proposed High-Speed A5/1 (HSA5/1)

In the proposed HSA5/1 cipher, a 36-bit keystream is generated in one clock cycle and is used to encrypt the full 16-bit data block and other parts of the ECS packet ahead of packet transmission. To achieve this, HSA5/1 makes use of eight registers, X, Y, Z, K_1, K_2, K_3, K_4, and ID. The registers X, Y, and Z are 41-bit, 43-bit, and 44-bit wide, respectively, whereas all other registers are 4-bit wide. All the registers are initially filled using a 148-bit secret key, K. The properties of HSA5/1 are shown in Table 7.1 and contrasted with those of A5/1. The 36-bit keystream generated at each transaction (transmission or reception) is comprised of three smaller keystreams, S_1, S_2, and ID. These keystreams are 16-bit, 16-bit, and 4-bit wide, respectively. The keystream S_2 is further divided into four 4-bit keystreams, K_1, K_2, K_3, and K_4. Please note that we have abused notation and used the same names for the 4-bit K_i registers and the 4-bit K_i keystreams as we use the register contents directly as the keystreams. On the other hand, the registers X, Y, and Z are used to update the S_1, S_2 and ID keystreams.

The generation of the $S_1 = \{s_0, s_1, \ldots s_{15}\}$ keystream differs from that used in A5/1 as follows:

1. The majority function and the keystream generation equations are as given in Table 7.1.
2. The 16 bits of S_1 are generated using the equation given in Table 7.1 where i_S, i_X, i_Y, and i_Z are the indices given in Table 7.2.
3. Depending on the data sent or received in the previous transaction, the shift registers are swapped according the 8th bit of the previous data stream. If the 8th bit is ON, the shift registers are swapped as follows:

$$x_1 : x_{40} = y_1 : y_{40} \tag{7.1}$$

$$y_1 : y_{40} = z_1 : z_{40} \tag{7.2}$$

Table 7.2 Index table for the generation of the 16-bit S keystream of Table 7.1

i_S	i_X	i_Y	i_Z	i_S	i_X	i_Y	i_Z	i_S	i_X	i_Y	i_Z
0	5	35	3	6	12	3	7	12	19	16	18
1	38	27	13	7	14	23	33	13	13	25	11
2	28	10	23	8	11	37	32	14	30	2	20
3	8	17	12	9	32	20	39	15	10	22	5
4	17	39	29	10	2	11	28				
5	25	33	24	11	18	36	34				

$$z_1 : z_{40} = x_1 : x_{40} \tag{7.3}$$

It must be noted that the 0th bit is not swapped. In any LFSR, the keystream of the original A5/1 will eventually repeat, albeit after a long period. On the other hand, the swapping introduced in HSA5/1 helps avoid such repetition and adds more entropy in the keystream generation. Another important consideration is that beside using a data bot for register swapping, other options are available for adding dependencies and increasing key generation entropy. For instance, the dynamic flags of the ECS packet may be used to control the swapping of the shift registers.

As for the $S_2 = K_1 \| K_2 \| K_3 \| K_4$ and ID keystreams,[1] they are generated as follows:

$$K_1 = K_1 \oplus \{ D_p[5], \ y_1, \ z_{19}, \ x_{37} \} \tag{7.4}$$

$$K_2 = K_2 \oplus \{ D_p[7], \ y_7, \ z_9, \ x_{31} \} \tag{7.5}$$

$$K_3 = K_3 \oplus \{ y_{21}, \ z_{39}, \ x_{11}, \ D_p[5] \} \tag{7.6}$$

$$K_4 = K_4 \oplus \{ y_{13}, \ z_{13}, \ x_{21}, \ D_p[7] \} \tag{7.7}$$

$$ID = ID \oplus \{ D_p[3], \ y_{29}, \ z_{21}, \ x_9 \} \tag{7.8}$$

where D_p is the data block of the previous transaction. The keystream S_1 is used to encrypt the data block, while S_2 and ID are used to encrypt the ECS packet headers. The secure ECS protocol using the proposed HSA5/1 cipher is presented next.

7.2.2 Secure ECS Communication

The overall secure ECS transmission flow is shown in Fig. 7.3, where two nodes are communicating using a single-channel link. In reference to Fig. 7.4, the transmission steps are as follows:

T1: Generate the keystreams S_1, S_2, and ID using HSA5/1 as described in the previous subsection.
T2: Integrate the 4-bit keystream ID with the 12-bit data block D, as shown in Fig. 7.4, to generate 16-bit packet data PD.
T3: Encrypt PD using the S_1 keystream to generate the encrypted packet data $EPD = \mathcal{E}(PD, S_1)$, where $\mathcal{E}$ is the HS5/1 encryption operator.
T4: Encode EPD using the ECS encoder. The ECS encoder will maximize the data rate and generates an ECS packet $CEPD = \mathfrak{C}(EPD)$, where $\mathfrak{C}$ is the ECS encoding operator. The ECS encoder will generate two 8-bit encoded

---

[1] The notation $\|$ is used for the concatenation operator.

data segments according to $\mathfrak{C}(EPD_1 \| EPD_2) = CEPD_1 \| CEPD_2$, with the corresponding NOI_i and $FLAG_i$ for $CEPD_i, i = 1, 2$.

T5: Encrypt NOI_i and $FLAG_i, i = 1, 2$, using the $S_2 = K_1 \| K_2 \| K_3 \| K_4$ keystream as follows:

$$ENOI_1 = K_1 \oplus NOI_1 \tag{7.9}$$

$$EFLAG_1 = K_2 \oplus FLAG_1 \tag{7.10}$$

$$ENOI_2 = K_3 \oplus NOI_2 \tag{7.11}$$

$$EFLAG_2 = K_4 \oplus FLAG_2 \tag{7.12}$$

T6: Transmit $CEPD = CEPD_1 \| CEPD_2$ in the ECS packet, as shown in Fig. 7.6, using the original NOI_i and $FLAG_i, i = 1, 2$. It is important to note that the original NOI_i and $FLAG_i$ are only used to transmit $CEPD_i, i = 1, 2$, but are not transmitted themselves. Instead, the encrypted versions $ENOI_i$ and $EFLAG_i, i = 1, 2$, are transmitted within the ECS packet.

T7: Finally, update S_1, S_2, and ID.

At the receiver, the decoding of secure ECS process is the reverse flow of the encoding, and the steps are as follows (Fig. 7.3):

R1: Receive $ENOI_1$ and $EFLAG_1$ of the first segment, and decrypt them using the K_1 and K_2 keys to recover the original NOI_1 and $FLAG_1$.

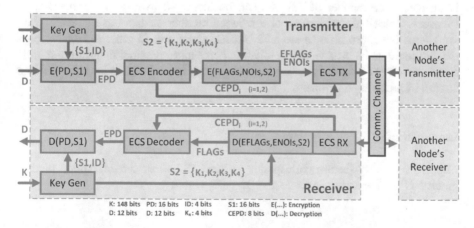

Fig. 7.3 Block diagram of a secure ECS transceiver. Note the symmetric multi-key generation. Note also the presence of two encryption blocks, one for data, $E(PD, S1)$, and one of signaling parameters, $E(FLAGs, NOIs, S2)$. Finally, note data authentication using the ID key

Fig. 7.4 ID integration into data

R2:   Use NOI_1 to log each ON bit index pulse series for the first segment and form
      $CEPD_1$.
R3:   Repeat steps R1 and R2 to receive NOI_2, $FLAG_2$, and $CEPD_2$ for the second
      segment.
R4:   Decode $CEPD_i$ to form EPD_i, $i = 1, 2$.
R5:   Decrypt $EPD = EPD_1 \| EPD_2$ using S_1 to recover PD.
R6:   Split PD into ID and data D.
R7:   If the decoded ID matches receiver's own ID, update keystreams S_1, S_2, and
      ID

Based on the above transmit and receive protocols, we can relate the three major
security features of authentication, confidentiality, and confusion to the steps of
secure ECS communication using HSA5/1 as follows:

Authentication

This ECS security feature corresponds to Step T2, where the ID bitstream bits
are equally distributed over the data segments as shown in Fig. 7.4. The receiver
recovers the ID bits only if the ECS packet is successfully decrypted using $\mathcal{E}^{-1}$
and decoded using $\mathfrak{C}^{-1}$. Let ID_T be the ID bits extracted by the receiver from the
transmitted ECS packet and let ID_R the ID bits generated by the receiver using
the symmetric keystream generator. In case authentication failure, $ID_T \neq ID_R$, the
ECS packet is discarded, and none of the $S1$, $S2$, K_i, $i = 1 \ldots 4$, and ID keystreams
is updated. The number of ID bits can be increased to improve authentication
robustness, albeit at the expense of a reduced number of data bits in each packet.
An alternative approach is to use a 32-bit ECS dataframe in order to increase both
the ID and data block size. Authentication is meant to preempt man-in-the-middle
attacks where the attacker's main goal is to use any invalid data to interfere with the
control of receiving equipment.

Confidentiality

This ECS security feature corresponds to Step T3, where the encryption of data
block PD is executed to generate EPD.

Confusion

This ECS security feature corresponds to Steps T5 and T6 and contributes to
confidentiality and integrity by making it difficult for an attacker to decrypt and
decode the ECS packet successfully. For each data block segment, the ECS encoder
generates NOI and $FLAG$ headers that are transmitted ahead of the segment
indices. Decrypting NOI and $FLAG$ correctly is an absolute pre-requisite to

successfully receive and decode the pulse streams and extract the transmitted data block. If NOI and $FLAG$ are compromised, the pulse counts will be compromised as well, and a transmission failure will occur, resulting in the receiver rejecting the ECS packet. A couple of transmission failure scenarios will be illustrated in the example of Sect. 7.3. In Steps T5 & T6, the NOI and $FLAG$ headers are replaced by their encrypted counterparts, which confound the attacker but provide the intended receiver with the means to recover the transmitted information. Encrypting NOI and $FLAG$ adds an extra shield of packet data protection that prevents the attacker from even "opening" the packet.

7.2.3   Multiple Layers of Security

For an attack to succeed, the ECS packet should first be received successfully. However, since the $NOIs$ of all the segments are encrypted, such reception is not likely. In the unlikely event that the attacker does receive a valid ECS packet, (s)he will be presented with the challenge of decrypting $FLAGs$. If $FLAGs$ are not decrypted correctly, the data block is unlikely to be decoded correctly. In the unlikely event that $FLAGs$ are decrypted correctly, (s)hc will be presented with the challenge of data block decryption. Furthermore, the ID of the transmitter needs to be authenticated. In summary and as shown in Fig. 7.5a, $NOIs$, $FLAGs$, data, and ID present four defense layers that an attacker should overcome to be able to decrypt the secure ECS packet correctly. A probabilistic analysis of the likelihood of a successful attack on secure ECS will be given in Sect. 7.4. For the sake of comparison, consider one of the common single-channel communication protocols, such as UART or 1-Wire, and assume it is used in conjunction with the A5/1 cipher. Such arrangement for secure, single-channel communication has only one security layer against which real-time attacks as described in [5] will be successful. The secure ECS case has the advantage of multiple security layers, as shown in Fig. 7.5.

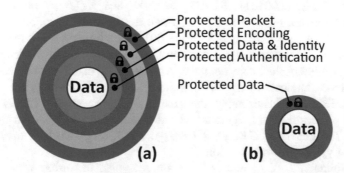

Fig. 7.5   Contrasting the security layers of two protocols: (**a**) ECS + HSA5/1 (this work) and (**b**) UART + A5/1

Furthermore, the single-clock cycle architecture of HSA5/1 allows us to more than double the key length without compromising the communication data rate.

7.3 Example of Secure ECS Communication

This section is devoted to a worked-out example to illustrates the inner working of secure ECS communication. The emphasis will be on highlighting the multiple security players that result from integrating the dynamic features of ECS with the high-speed HSA5/1 crypto. A secure ECS packet construction example is given in Fig. 7.6 where an ECS protocol having $B = 16$ and $l = 8$ is considered. Assume that the secure ECS system needs to transmit the decimal number 1384 (i.e., 010101101000 in binary). Assume further that the HSA5/1 key generator has already generated the keystreams $S1 = 1010100110001110$, $K_1 = 0010$, $K_2 = 0001$, $K_3 = 0111$ and $K_4 = 0011$, and the identification key $ID = 0101$. In the following subsections, we will first show how the secure ECS packet is formed and transmitted. We will then show how it is decrypted and decoded at the receiver.

7.3.1 Secure Packetization

The ID bits are added at both ends of the data block, as in Fig. 7.4, to form ECS data packet PD=**0101**010110100**0001**. The PD is then encrypted using the S1 key to generate the encrypted PD, EPD=1111110000101111. The process of generating EPD is shown in the "Data Encryption" box of Fig. 7.6.

In the next step, the EPD is used to apply ECS encoding, where EPD is broken into two independent segments to reduce the index values of the most significant bits. Both inversion and reversal operations are applied to the first segment to reduce both the number of ON bits and their index values. This results in the encoded segment with bits 00001011, $FLAG_1 = 0011$, and $NOI_1 = 0011$. Similarly, the second segment needs inversion only and results in the encoded segment bits 00000101, $FLAG_2 = 0010$, and $NOI_2 = 0010$. Recall that NOI is the size of the support subset of the segment, i.e., the number of ON bits, which is identical to the Hamming weight of the binary segment. Among all the data bits of both segments, only the index numbers of the ON bits are selected to form the "encoded EPD", denoted CEPD. For this example, the generated CEPD consists of index numbers 1, 2, and 4 from the first segment and index numbers 1 and 3 from the second segment. The generated $CEPD$, $FLAG_{1,2}$, and $NOI_{1,2}$ are then used for both packet encryption and transmission.

The original $FLAG_{1,2}$ and $NOI_{1,2}$ are encrypted to further secure the ECS packet as they are necessary for the successful decoding of the transmitted data. Packet encryption uses the keystreams, K_1, K_2, K_3 and K_4, to generate the encrypted $FLAG_{1,2}$, denoted $EFLAG_{1,2}$, and the encrypted $NOI_{1,2}$, denoted

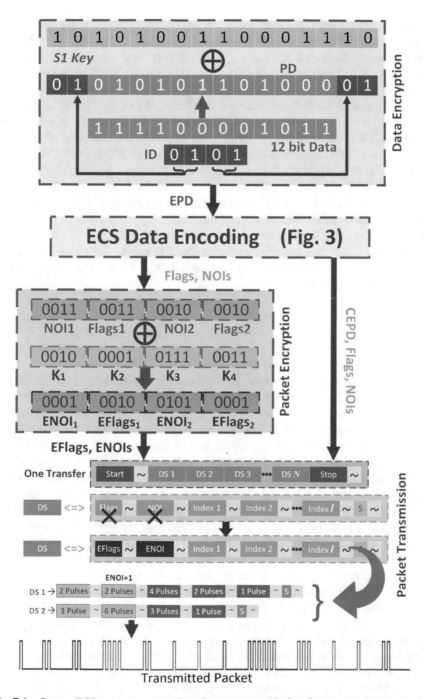

Fig. 7.6 Secure ECS packet construction: the top green block refers to the data encryption function $E(PD, S1)$ of Fig. 7.3 the middle gray block refers to the ECS encryption function $E(FLAGs, NOIs, S2)$ of Fig. 7.3. The bottom pulse stream is a combination of the pulses generated for the encrypted data and those generated (yellow block) for the encrypted ECS parameters

$ENOI_{1,2}$. The resulting values of $EFLAG_{1,2}$ and $ENOI_{1,2}$ for the current example are: $EFLAG_1 = 0010$, $ENOI_1 = 0001$, $EFLAG_2 = 0001$ and $ENOI_2 = 0101$. These $EFLAG_{1,2}$ and $ENOI_{1,2}$ are then fed into the ECS transmitter. The encryption of the ECS packet is shown in the "Packet Encryption" box of Fig. 7.6.

In reference to the "Packet Transmission" box of Fig. 7.6, the ECS packet is constructed using the $CEPD$, $EFLAF_{1,2}$, and $ENOI_{1,2}$. Note that encrypted $FLAG$ and NOI are transmitted rather than original ones. However, all the index numbers in $CEPD$ are transmitted using the original $NOI_{1,2}$ of $CEPD$. For example, the first segment of the $CEPD$ block includes 4 pulses for $FLAG_1$, 4 pulses for NOI_1 (3 and an additional pulse), and the 3 index numbers. On the other hand, the secure ECS packet includes 2 pulses for $EFLAG_1$, 2 pulses for $ENOI_1$ (1 and an additional pulse), and the 3 index numbers. Similarly, the second segment of the $CEPD$ block includes 2 pulses for $FLAG_2$, 3 pulses for NOI_2 (2 and an additional pulse), and the 2 index numbers. However, the secure ECS packet includes 1 pulse for $EFLAG_2$, 6 pulses for $ENOI_2$ (5 and an additional pulse), and the 2 index numbers.

7.3.2 Secure Reception

At Trudy's end, $ENO_{1,2}$ are used as received to decode the rest of the packet, which will result in transmission failure. This is because $ENOI_1$ leads her receiver to believe that one index number is being transmitted while, in fact, a total of 3 indices is being transmitted. Trudy's receiver will decode only the first index number of the first segment and will consider the rest of the pulse streams as originating from a second segment. As a result, the decoding error will cascade to the second ECS segment and from there to the subsequent secure ECS packets. This failed transmission scenario is illustrated in Fig. 7.7a and is contrasted with Bob's successful transmission. In Fig. 7.7b, another example of failed transmission is presented where Trudy's receiver expects three index numbers to be transmitted whereas the total number of transmitted indices is one. As a result, Trudy's receiver will consider the sync and $EFLAG_2$ pulses as index numbers, which in turn cascades the decoding error to the second segment for which Trudy's receiver will decode Index2 as the NOI_2 and keep waiting for the remainder of the pulse streams. As shown in Fig. 7.7c, the error will cascade to the subsequent ECS packets and compromises their decoding as well. It is conceivable that packet-to-packet decoding error detection schemes can be designed, but decoding error detection within a packet cannot be avoided in the absence of the correct keystreams.

For a transaction to be successful, both ends of the communication link must have the same keys. To successfully decrypt a received ECS packet, the keys are first used in $ENOI_{1,2}$ decryption so as to know the exact number of indices in the ECS packet. Then they are used in $EFLAG_{1,2}$ decryption to find out the ECS decoding scheme of the data bits. The process is explained graphically in Fig. 7.8 using a set

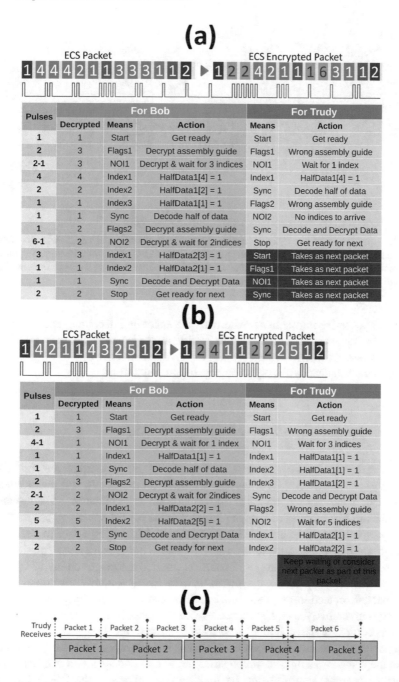

Fig. 7.7 Examples of erroneous packet decryption by an intruder: (**a**) Decryption failure due to wrong decryption of ECS parameters resulting in early packet termination and the erroneous start of a new-packet decryption; (**b**) Decryption failure due to wrong decryption of ECS parameters resulting in a wait state. In (**c**), the scenario of (**a**) is further illustrated as a cascade of decryption errors (red line) encompassing all the packets that follow the wrong decryption of Packet 1 (gray boxes)

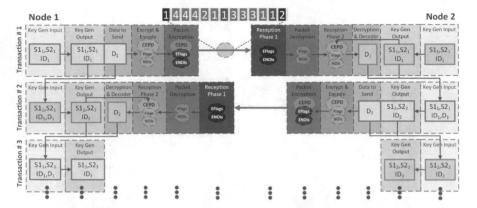

Fig. 7.8 Illustration of the symmetric and dynamic nature of secure ECS with the keys of Transaction#2 dependent on the data of Transaction#1. This figure further illustrates the layered architecture of secure ECS with decryption of the ECS parameters being a necessary condition for the decryption of the data

of three successive transactions, where all the encoding, decoding, encryption, and decryption steps are illustrated. Figure 7.8 illustrates how the inputs and outputs of the key generator change at each transaction so that both ends of the communication link are synchronized with respect to keystream generations.

7.4   Cryptanalysis of the Multilayer Cipher

A considerable amount of literature is available on attacks to break the A5/1 cipher such as guess and clock [3], time-memory trade-off attack [7], and linear equation solve [19]. Time-memory trade-off allows an attacker to reconstruct the key in one second from two minutes of known plaintext at the cost of expensive preprocessing of 2^{48} steps to compute around 300 GB of data. Time complexity to break A5/1 through solving linear equations is $2^{40.16}$. These complexities are for the 64-bit cipher. Not only does our proposed solution provide a stronger cipher using a 128-bit key, but also it presents the attacker with an even harder challenge through ECS packet protection and data block authentication. Aside from using the 128-bit key, our multilayer cipher increases the complexity by a factor of 2^{20}. This factor accounts for the additional keys used to encrypt $FLAG_{1,2}$ and $NOI_{1,2}$ of the ECS protocol. The modifications to A5/1 and its integration with the ECS protocol make the secure transmission much stronger. Not only does it add confusion to the cipher system but also diffusion by making the settings of the HSA5/1 registers dynamically dependent on the plaintext.

To reach the central HSA5/1 cipher system, an attacker has first to challenge herself to receive the packet successfully. The plaintext-dependent encrypted information for the number of support indices $(ENOI_{1,2})$ needs to be decrypted successfully before their decoding. Otherwise, packet failure will occur. Furthermore, the successful decryption of an encrypted encoding flag, $EFLAG_i$, depends on $ENOI_i$ itself, which provides additional protection to the decoding process. If an attacker is successful in guessing both of these, then data authentication using ID would be the next barrier. Recall that these three ECS packet fields are updated at each packet transmission along with the generated keystream. Moreover, like the keystream, these keys are plaintext-dependent, which makes it difficult for an attacker to send any invalid data to destroy synchronization or control the device. The probability that an attacker could guess a packet that is acceptable to the receiver is

$$P_{att.} = \frac{1}{2^{l_D} \times 2^{l_{ID}} \times 2^{4l_K}} = \frac{1}{2^{l_{PD}} \times 2^{4l_K}} \tag{7.13}$$

where l_D, l_{ID}, l_{PD}, and l_K are the lengths of input data block, ID, ECS data block, and the K_i keystreams, respectively. In our example, where we have used 12-bit D, 4-bit ID, and 4-bit l_K, the probability is $\approx 0.23 \times 10^{-9}$. This is because an attacker has to guess two 4-bit NOI, two 4-bit $FLAG$, one 4-bit ID, and a 12-bit D to form a valid ECS packet.

Another issue that could arise is when we have constant data, for instance, when the sensor node is powered off and the digital value is all zeros. In this case, HSA5/1 reduces to A5/1 and can, therefore, be broken as long as the A5/1 key is guessed correctly. Again, to reach this point, successful packet reception and its decoding are necessary. To further mitigate this issue, we recommend using the HSA5/1 register contents after N cycles by discarding the initial 256-bits of s. In case an attacker wants to store a transmitted packet and then applies exhaustive search to find the key, (s)he has two main challenges. First, if we suppose that she can recover the key, she must know the previously transmitted data as the keystream and register contents are data-dependent. In most cases, previously transmitted data is not known to her. Second, even in case such data is known to the attacker, (s)he must adopt a brute-force exhaustive search method with insurmountable complexity given

$$W_{exh} = \frac{2^{l_S} \times 2^{l_{ID}} \times 2^{4l_K}}{2} \tag{7.14}$$

where l_S is the length of the HSA5/1 key. We have used $l_S = 128$, which is twice the standard $A5/1$ key. The brute-force attack complexity on secure ECS is 2^{147}.

7.5   Implementation, Cipher Overhead, and Comparison with Prior Art

Several hardware platforms have been used to implement secure ECS, including a low-end microcontroller, a high-end FPGA, and an ASIC platform. Both the microcontroller and FPGA platforms have been prototyped in hardware, while the ASIC platform has been evaluated using standard logic-synthesis tools.

7.5.1   Microcontroller Prototype

In this implementation, we have used a low-end microcontroller, MKL25Z, mounted on an FRDMKL25Z board from Freescale. We have implemented the full secure ECS system in C that includes the ECS communication, a keystream generator, encryption/decryption, and an authentication flow. There are two nodes in the experimental setup that establish a secure communication link using the ECS technique. The setup transmits data securely from one node to the other, and the other node sends received data back to the first node to complete a full transmission link. Received and transmitted data are then compared to verify two-way communication.

7.5.2   FPGA Prototype

In this implementation, we have used the Xilinx Virtex-7 FPGA platform to run and verify secure ECS using the Verilog Hardware Description Language. We have also set up a network of devices in a master–slave configuration to mimic an ECS-based IoT network in support of a smart building use case. Each link of the network has a unique set of keys that is used for communication between the nodes of the link. Each node in the network maintains a key table for all its links, as shown in Fig. 7.9a. The Master device starts communication with a particular node fetching the corresponding link key from the key table. Both link nodes update their keys if the transaction is successful (i.e., legitimate response from the slave). Though we have used a master–slave configuration in our experiments, other network configurations are readily conceivable, as shown in Fig. 7.9b, where the yellow box attached to each node represents the key table. Rigorous simulation and hardware experimentation are performed for functional testing and verification of the implemented systems. The experiments confirm that secure ECS transmission works flawlessly. It is important to note that our proposed multilayer cipher system is not a multiple, nested, or cascade cipher [13]. In the latter cases, the same data is encrypted multiple times using the same cipher but with different keys or using different ciphers (e.g., AES and RSA). In our multilayer cipher, the data is encrypted once in one layer, and the flags used to encode the data as pulses are encrypted, also once, in a different layer. It is of course conceivable to extend our proposed multilayer cipher to a multiple, cascaded cipher, but this is outside the scope of this chapter. Our testing and validation method is unique to secure ECS, and methods based on cascaded ciphers cannot be used.

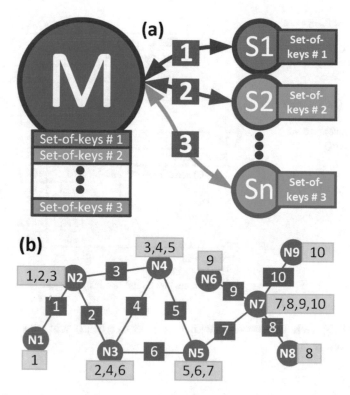

Fig. 7.9 Secure ECS networking options: (**a**) Master–slave configuration (**b**) Other possible configurations with the blue circles showing the ECS nodes. Each node has a yellow box attached showing its secure ECS routing links

7.5.3 ASIC Synthesis

Along with the embedded and FPGA prototypes, we have also synthesized the secure ECS protocol using a Synopsis logic-synthesis flow and a GLOBAL-FOUNDRIES 65 nm technology node. We have found that the system shown in Fig. 7.3 (without the ECS codec) maintains the low-power operation of ECS consuming only 27 μW with a gate count of $\approx$2780 at a clock frequency of 25 MHz. The synthesis results are shown in Table 7.3. Additionally, even with the security overhead, the secure ECS system maintains the low-power operation of ECS and is still of lower power consumption as compared to CDR-based serial transfers. Such a comparison is given in Table 7.4. Please note that the schemes presented in the five references in Table 7.4 are for state-of-the-art low-power CDR schemes *without any crypto overhead*. They can be used with the NRZ serial transfer for data transmission over a single channel. Our proposed multilayer security architecture's low-power efficiency is showcased in Table 7.4, which compares the state-of-

Table 7.3 ASIC synthesis results in GLOBALFOUNDRIES 65 nm technology

	Power (μW)	Area (*gate count*)
ECS Crypto	27	$\approx$2780
ECS Codec	26.6	2356
Total	53.6	$\approx$5136

Table 7.4 Comparisons with serial communication based on clock and data recovery (CDR)

	Power (uW)			
	SRL[a]	CDR	Total[b]	
ECS	26.6	N/A	26.6	Chapter 2
Secure ECS[c]	53.6	N/A	53.6	
Normal serial transfer	32.1	70	102.1	[37]
		62.5	94.6	[38]
		90	122.1	[12]
		57.5	89.6	[77]
		60.6	92.7	[73]

[a]Serializer
[b]SRL+CDR
[c]Keystream generation in one clock cycle

the-art NRZ+CDR transceivers (without crypto overhead) with our secure ECS architecture.

7.5.4 Secure ECS Design Alternatives

The results in Table 7.4 are obtained using the secure ECS architecture under the assumption that only one clock cycle is available to generate all the keystreams. However, when more than one clock cycles are available for cryptographic operations, the architecture can be readily modified to generate keystreams in as many clock cycles as available. It is anticipated that such change would decrease power consumption and hardware resources but at the expense of a lower data rate. There are two design alternatives for adapting the architecture to the number of clock cycles available for crypto processing.

The first is to reduce the lengths of the registers such that fewer bits of the keystream (e.g., 4 bits) are generated in each clock cycle and eventually concatenated to form a 16-bit key. However, this approach is not recommended because of the need to generate simultaneously one main keystream and five sub-keystreams. Indeed, decreasing the length of the registers will compromise the strength of the cipher and make it vulnerable to attacks.

The second design alternative is to completely update the registers during the first clock cycle and generate a portion of the keystream bits. The hardware can be re-used to generate the rest of the bits in subsequent clock cycles without changing the initial state of the registers. This modification would result in a lower data rate but with fewer gates involved in keystream generation, which would reduce power

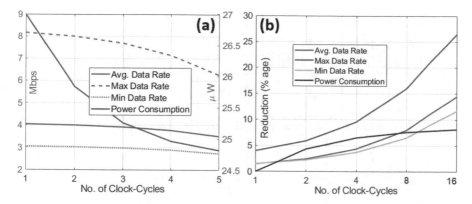

Fig. 7.10 (**a**) Trade-off between power consumption and data rate as function of crypto latency measured in clock cycles. Note that a single-clock-cycle crypto latency results in the highest power consumption. (**b**) Impact of crypto latency on data rate. Plots (**a**) and (**b**) indicate that for this implementation, a crypto latency of 3 clock cycles provide the best trade-off between power reduction and data rate maximization

consumption. The overall impact and the trade-off between power consumption and data rate as a function of the crypto clock cycles are shown in Fig. 7.10a. The trade-off is further clarified in Fig. 7.10b, where relative reductions in power and data rate are plotted vs. the available crypto clock cycles. Note that for the same number of crypto clock cycles, the relative reduction in data rate is much higher than the relative reduction in power, with the difference becoming more pronounced as the number of crypto clock cycles is increased. It is therefore concluded that the crypto overhead in ECS is essentially driven by the data rate with maximum data rate achieved using a single-clock-cycle crypto implementation.

Finally, we would like to point out the secure ECS design alternative based on the domain-specific instruction set architecture, ECSIA, highlighted in Chap. 8. A crypto ISA extension along with the supporting HSA5/1 crypto block can be added to ECSIA that combines both the flexibility of a securely programmable IoT communication interface with the low-power, high data rate features of ECS.

7.5.5 Comparison with Lightweight Ciphers

Finally, HSA5/1 and the secure ECS transceiver are compared with several lightweight ciphers from the prior art. Table 7.5 summarizes the comparison results which clearly illustrate the competitive performance of the standalone HSA5/1 cipher with respect to other published lightweight ciphers. The more interesting comparison is the one involving the full secure transceiver, which calls the following important remarks:

1. *The HS5/1 cipher*: Given its single-clock cycle design, the 128-bit HS5/1 cipher achieves the smallest latency of 1 clock cycle (similar to the 80-bit Trivium

Table 7.5 Comparison of secure ECS with other lightweight ciphers. Clock frequency 25 MHz

Cipher	Mode (E,D,E$_c$)[a]	Key (bits)	Clock cycles	Area (GE)	Power (μW)	DR[b] (Mb/s)	TP[c] (Mb/s)	EO/ PE[d]	DyDR/ DTP[e]	Technology (nm)
PRESENT [63]	E	128	559	1391	N/A	N/A	2.88	NO	NO	180
CLEFIA [1]	E+D	128	176	2996	N/A	N/A	18	NO	NO	130
AES [14]	E+D	128	1032	3400	N/A	N/A	3.13	NO	NO	350
DES [63]	E+D	56	144	2309	N/A	N/A	11.1	NO	NO	180
SEA [40]	E+D	96	93	3758	N/A	N/A	25.8	NO	NO	130
ICEBERG [40]	E+D	128	16	7732	N/A	N/A	100	NO	NO	130
HIGHT [24]	E+D	128	34	3048	N/A	N/A	47	NO	NO	250
Trivium [20]	E+D	80	1	2599	N/A	N/A	25	NO	NO	130
Grain [20]	E+D	80	1	1294	N/A	N/A	25	NO	NO	130
Simon [4, 6, 32]	E+D	128	>116	3603	204	N/A	45.5	NO	NO	65
Midori [4]	E+D	128	>20	3959	152	N/A	100	NO	NO	65
Sepck [6, 32]	E	128	>64	2727	129	N/A	94	NO	NO	130
HSA5/1 (this work)	E+D	128	1	2780	27	900	900	NO	NO	65
Secure ECS (this work)	E+D+E$_c$	148	47–129	5136	53.6	3.1–8.5	3.1–8.5	YES	YES	65
(TxRx+HSA5/1)			(98)[f]			(4.1)[f]	(4.1)[f]			

[a]E: Encoder, D: Decoder, E$_c$: ECS Signaling

[b]Data Rate

[c]Throughput

[d]EO: Encoding Obfuscation, PE: Packet Encryption

[e]DyDR: Dynamic Data Rate, DTP: Dynamic Throughput

[f] Average

and Grain). Among the 128-bit ciphers with both encoding and decoding implemented in hardware, it has the smallest footprint and, by far, the largest throughput.

2. *Footprint of the secure ECS transceiver*: The full secure transceiver has a footprint measured in GE that is quite reasonable and is interestingly less than the ICEBERG cipher alone.

3. *Cipher overhead:* The HS5/1 cipher has essentially doubled the footprint and power consumption of the ECS transceiver, yet the total amount of area and power consumed by the secure ECS transceiver are well within the envelope of a constrained IoT node.

4. *Impact of transceiver design:* The last two lines of Table 7.5 are meant to illustrate the interplay between cipher and transceiver design. While the standalone cipher has a data rate and throughput of 900 MB/s at a clock rate of 25 MHz, such data rate and throughput cannot be sustained by the transceiver due to the specifications of the signaling protocol. It would be interesting to explore such interplay between the security component and the communication component for the prior standalone ciphers.

5. *Multilayer cipher:* The multilayer nature of the secure ECS cipher is evident in the number of key lengths used at the transceiver level (148 bits) vs. the key length used by the cipher itself (128 bits). The additional 20 bits are used for the encryption of packet parameters and the obfuscation of encoding flags. This makes the secure ECS transceiver more challenging to attack than HS5/1 itself or any of the lightweight ciphers referenced in Table 7.5.

7.6 Conclusions

The proposed secure, single-channel communication system exploits the unique features of the ECS protocol and adds multiple layers of security to the transmission with low impact on ECS data rate and power performance. It does so while presenting a potential attacker with a layered set of hard-to-solve challenges. The core of the secure ECS system is HSA5/1, a novel, strengthened, low-latency architecture the A5/1 keystream cipher that uses a total key length of 148-bit, out of which 128-bits are used for the HSA5/1 keystream generation and 20-bits are interleaved with the ECS protocol to provide additional layers of packet encryption. The secure ECS solution features both confusion and diffusion defenses while providing both data confidentiality and packet authentication. The secure ECS system has been prototyped on embedded microcontroller, FPGA, and ASIC platforms with all the prototypes confirming the low overhead of the HSA5/1 crypto block. The secure ECS prototypes have achieved data rate, power consumption, small footprint figures that are all in line with the original ECS attributes. Conceivably, HSA5/1 may be used with other single-channel protocols albeit without the additional crypto layers that are tied up to the ECS packet specification.

This chapter highlights the advantages of a tight integration between the IoT communication protocol on the one hand and the lightweight cryptographic algorithm on the other. As illustrated in the integration of ECS with HSA5/1, this integration has resulted in a secure communication protocol that is very well adapted to constrained IoT nodes. This {ECS,HSA5/1} case will serve as a reference point in our future work which will include more extensive evaluations of, and comparisons with, other lightweight cipher algorithms, especially in terms of their compatibility with the multilayer cryptographic architecture presented in this chapter.

Chapter 8
Domain-Specific ECS Processor

> *In view of these inevitable limitation, architects now widely believe that the only path left for major improvements in performance-cost-energy is domain-specific-architectures (DSAs).*
>
> David Patterson

This chapter introduces the domain-specific architecture (DSA) of a novel streaming processor dedicated to the edge-coded signaling (ECS) communication. The presentation of this DSA will be self-contained in that it will include the instruction set architecture (ISA) itself, the minimal micro-architecture needed to implement it, and an analysis of its performance using the thin software stack needed to run programs on the implemented processor. In particular, the micro-architectural opportunities for designing a minimal, area- and power-efficient processor for this important domain will be highlighted. Metrics such as data rate, energy efficiency, area, and power will be given and compared with those of various hardwired implementations of the IoT signaling techniques under consideration.

8.1 Introduction

DSAs are computing platforms that are tailored to the characteristics of a well-defined domain. They differ from general purpose architectures in that they are designed to execute a specific set of tasks extremely well. They also differ from Application-Specific Integrated Circuits (ASIC) in that they are programmable to efficiently work on several applications in a given domain. A DSA comes with its domain-specific instruction set, which constitutes the core of its programmability. One important benefit that a DSA has over a general purpose CPU is that its task-tailored programs have significantly fewer instructions than those of a CPU and, therefore, achieve significant performance gains and power savings due to a much lower instructions-per-program metric.

© Springer Nature Switzerland AG 2022
S. Muzaffar, I. M. Elfadel, *Secure, Low-Power IoT Communication Using Edge-Coded Signaling*, https://doi.org/10.1007/978-3-030-95914-2_8

Another important DSA benefit is that it offers micro-architects with many additional opportunities to improve hardware performance above and beyond instruction-level parallelism or hardware, e.g., multi-core, parallelism. These opportunities are due to the domain knowledge that is captured in the domain-specific instruction set. Because of these new hardware opportunities for micro-architects, DSAs have been declared as the "only path left" for a hardware-centric future in computer architecture [22, 62].

The main elements of a successful DSA are the following:

1. *Domain-specific instruction set architecture*: A new instruction set has to be designed almost from scratch to embody domain knowledge. Currently, popular domains include Artificial Intelligence (AI), bio-informatics, the Internet of Things, and crypto-currencies. Traditional domains include graphics, digital signal processors, and cryptographic processors.
2. *Domain-specific software stack*: A new software stack has to be created in support of the domain-specific ISA. The stack will comprise compilers, debuggers, profilers, software development kits (SDKs), and application-programming interfaces. A very good example of such a stack is NVIDIA's CUDA environment for its GPUs.
3. *Domain-specific performance metrics*: Every knowledge domain will introduce its own task-specific performance metrics to evaluate various implementations of a domain-specific ISA. Examples of such new domain-specific metrics include the vertex-shader duration (VS) in graphics and the number of multiply accumulates (MACs) in artificial neural networks.

Two recent instances of a successful DSA are Google's Tensor Processing Unit (TPU) [28, 29] and IBM's AI core chip [15] for the acceleration of machine learning tasks. Each of these accelerators has its own custom instruction set along with a hardware micro-architecture meant to achieve maximum performance. As compared to CPUs and GPUs developed in similar semiconductor technologies, machine learning inference on Google's TPU is 15 to 30 times faster and is 30 to 80 times better in energy efficiency. The IBM AI core has achieved more than 90% sustained utilization across a range of neural network topologies. It is important to stress that these processors are domain-specific in that they can be programmed to accelerate not just deep convolutional neural networks as in [75] or [35] but also other machine learning workloads such as multilayer perceptrons (MLPs) or long-short-term-memory (LSTM) networks.

Incidentally, the TPU and the AI core are meant to accelerate the execution of workloads arising in the context of *artificial* neural networks and are therefore to be distinguished from brain-inspired, *neuromorphic* platforms such as [18, 70] and [9], whose computational model is that of the spiking neural activity of the nervous system.

Recent examples of programmable accelerators in application domains other than neural networks or neuromorphic computing include the cryptographic processor reported in [21], the biomedical platform processor [33], and the string matching accelerator [39]. In the area of network communication, it is worth mentioning

Cisco's routers where the main CPU (e.g., MPC860 PowerQUICC processor from Motorola/NXP) includes an on-chip Communication Processor Module (CPM) [17, 36, 43]. The CPM is a RISC microcontroller dedicated to several special purpose tasks such as signal processing, baud rate generation, and direct memory access (DMA). CPM may, therefore, be considered a domain-specific processor for the network-routing domain.

This chapter introduces the DSA of a novel streaming processor for an application domain that has not been addressed in the open literature, namely, the domain of IoT communication. The presentation of this DSA will be self-contained in that it will include the ISA itself, the minimal micro-architecture needed to implement it, and an analysis of its performance using the thin software stack needed to run programs on the implemented processor. In particular, the micro-architectural opportunities for designing a minimal, area- and power-efficient processor for this important domain will be highlighted. Metrics such as data rate, energy efficiency, area, and power will be given and compared with those of various hardwired implementations of the IoT signaling techniques under consideration.

In the specific domain of communication protocols, there have been two main approaches for their implementation. In the first approach, the communication engineer programs the entirety of the protocol on a microprocessor and controls their selection and parameters through registers. This is a standard practice that is followed for data transfer protocols such as I^2C, I^2S, SPI, UART, and CAN [11]. The second approach is to design an ASIC for the newest generation of the protocol and make it backward compatible with older versions as in the case of USB 2.0 and 3.0 [76]. The latter approach increases silicon area and power consumption and does not provide any customization features. Clearly, a pure software implementation, e.g., assembly language, on a general purpose processor often leads to an inordinate number of instructions to execute, which results in a drastic reduction in achievable data rates. It is one of the major benefits of a DSA to reduce the number of instructions needed to run a specific task, while its programmability enables the economic implementation of several variants of a given task.

The novel DSA introduced in this chapter is dedicated to ECS family of communication protocols. In line with the original protocol, the DSA architecture is called "Edge-Coded Signaling Interface Architecture" (ECSIA). The ECSIA processor, whose main components are shown in Fig. 8.1, possesses the following important properties:

1. Its ISA is a compact RISC-like architecture with 22 instructions, only one of which has a branching condition.
2. It is implemented as a full processor with its own instruction memory, and instruction fetching and decoding.
3. It is a streaming processor with no data memory. Incoming data is streamed directly into the register file.
4. It is a secure processor in that a cryptographic block can be added to the micro-architecture and enabled using an additional instruction in the ISA.

Fig. 8.1 Main components
of the ECSIA
micro-architecture

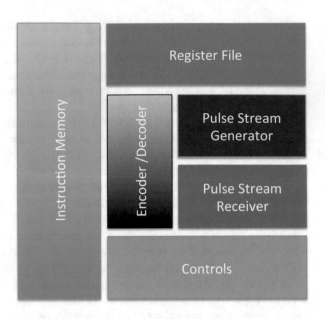

5. It has a thin software stack that enables the coding and compiling of ultra-compact C programs for implementing the various ECS protocols.
6. It brings power consumption closer to a dedicated ASIC design by enabling the sharing of optimally designed common and dedicated hardware modules and by providing the flexibility to reconfigure information flow between the various modules.
7. It enables the customized implementation not only of the standard ECS protocols but also of any communication protocol that uses the same underlying idea of transmitting information in the form of pulses.

8.2   Edge-Coded Signaling Interface Architecture (ECSIA)

As is clear from Chap. 2, ECS family members share the same functional and structural features. The functional features include segmentation and encoding, while the structural features include using *Flags*, *NOS*, and inter-symbol separators in the ECS packet. The ECS family members further share the same receiver mechanism, which consists of simply detecting and counting the edges of the incoming pulses. Their packet formats are also quite similar. While similar in structure and functions, they also leave a significant room for variations on the common themes of segmentation, encoding, representation, packetization, and pulse transmission. Possible variations may include the transmission of both the index numbers and decimal values, which is a form of redundancy to improve transmission reliability. Another form of redundancy is to repeat the transmission automatically with only a single start signal. To support the existing protocols and any possible variations within the ECS family, this chapter introduces a domain-specific instruction set, the ECSIA, to bring the various edge-coded signaling techniques under one umbrella. ECSIA can be used to generate not only the standard protocols with tunable communication parameters (i.e., segment size, inter-symbol separations, pulse width) but also it can be used to develop other customized communication techniques that use the same underlying idea of transmitting information in the form of pulses. The ECSIA micro-architecture uses a set of fundamental, optimized, hardware building blocks that can be programmed to implement various versions of ECS. These building blocks include pulse generators, ON bit counters, adders, segmenters, encoders, register files, and memories. In this chapter, the ECSIA processor is considered a self-contained, stand-alone streaming processor. However, it can also be embedded as a co-processor in a system-on-chip (SoC) along with other CPUs, GPUs and accelerators. The logic circuit details of the micro-architectural building blocks are given in Sect. 8.3. The rest of this section will be devoted to highlight the ISA itself.

8.2.1 Register Set

To enable protocol customization, registers are needed to enable the configuration of hardware modules, the control of data flow, the buffering of input and output data, and the tracking of execution status. To accommodate these needs, the ECSIA uses three types of registers:

1. General purpose registers: They form a set of eight 8-bit registers, R_0 through R_7, which are programmer-accessible.
2. Control registers: There are two 8-bit registers, $Ctrl_0$ and $Ctrl_1$, which are used to store protocol configuration parameters such as mode of transaction (transmitter or receiver), segment number, segment size, and pulse width in terms of a number of clock cycles. These control registers are initially set by the programmer through specific instructions, but once set, they become accessible only to the system.
3. IO register: This is the *LoadReg* register, which is a 16-bit, I/O-dedicated register used to read the I/O port, set the I/O port, and store the updated results after an instruction is executed. Like the control registers, *LoadReg* is a privileged register and is accessible only to the system.

These register types are summarized in Table 8.1. In the remainder of this chapter, the word register will solely refer to a general purpose register.

8.2.2 Instruction Formats and Types

The ECSIA instruction formats are shown in Fig. 8.2, and the assembly language instructions are given in Tables 8.2 and 8.3. The ECSIA instructions are all 16 bit

Table 8.1 ECSIA register set

	Register	Type	Organization
1	R_0–R_7	8-bit GP[a]	8-bit value
2	$Ctrl_0$	8-bit SP[b]	[0, Mode, 3-bit SegNum, 3-bit SegSize]
3	$Ctrl_1$	8-bit SP	8-bit pulse width
4	LoadReg	16-bit SP	16-bit value

[a] General purpose
[b] Special purpose

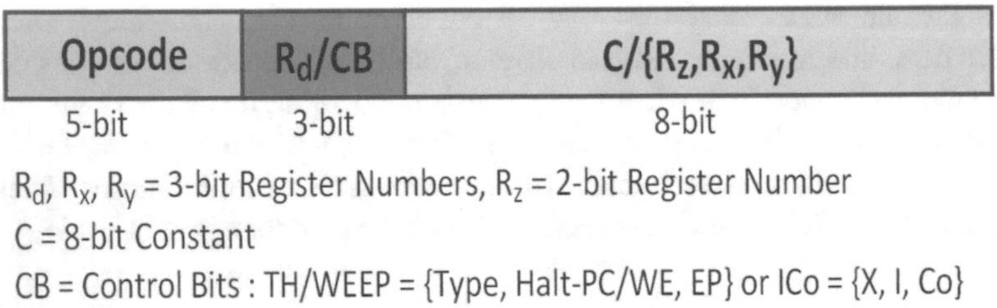

Opcode	R_d/CB	C/{R_z,R_x,R_y}
5-bit	3-bit	8-bit

R_d, R_x, R_y = 3-bit Register Numbers, R_z = 2-bit Register Number

C = 8-bit Constant

CB = Control Bits : TH/WEEP = {Type, Halt-PC/WE, EP} or ICo = {X, I, Co}

Fig. 8.2 ECSIA instruction formats

long in line with a RISC architecture. The 16-bit length is chosen to accommodate ECS protocols based on 16-bit data segments. Each instruction is divided into three main parts: Opcode, R_d/CB, and $C/\{R_d, R_x, R_y\}$. The 5-bit *Opcode* represents the type of operation. R_d, R_x, and R_y represent 3-bit register numbers. R_z represents a 2-bit register number. C is the constant operand value. The CB field contains the control bits and, depending on the instruction, can be either $TH/WEEP$ or ICo. The $TH/WEEP$ is the combination of three bits representing the *type of operand (T)*, *Halt-PC (H)* or *Write enable (WE)*, and *extra pulse enable (EP)*. The ICo is the combination of I and Co bits that are used to specify the condition for copying the register contents. In view of these two CB combinations, the instructions can be grouped into three types: one that uses $TH/WEEP$ bits (highlighted in gray in Table 8.2), a second that replaces $TH/WEEP$ with ICo bits, and a third that does not care about the CB field (highlighted in dark gray).

Type I

The first type of instructions, highlighted in gray, handles one operand at a time and is used in operations such as to read/write the I/O port, set/clear the *LoadReg*, set various communication protocol parameters, and send/receive pulse streams. These instructions use CB in the second part of the instruction format where CB represents $TH/WEEP$. T is used to set the type of operand (register or a constant) in an instruction. H/WE is used either to halt the PC during the transmission of pulse streams or to enable the store operation of received pulse count to a specified register. The bit EP is used if an extra pulse should be added to the transmitted pulse

Table 8.2 ECSIA instruction set

	Instruction	CB	Description	Example
1	RP	–	Load data from input pins to data register.	RP
2	WP	–	Output the received data from data register to the pins.	WP
3	SSS C	TH/WEEP	Set segment size (C = 0,1,2 for 4 bits, 8 bits, 16 bits).	SSS 1
4	SSN C	TH/WEEP	Select segment number (C = 0,1,2,3).	SSN 2
5	SM C	TH/WEEP	Set mode (C = 0,1 for transmitter, receiver). Setting RX mode clears LoadReg, and setting TX loads input into LoadReg.	SM 0
6	SW C	TH/WEEP	Set the width of pulse (C = integer specifying cycle count).	SW 2
7	IV R_x,R_y	ICo	Inverse the selected segment. R_x=NOI & R_y=Flags (R_x/R_y= R_0,R_1,...R_7).	IV R0,R1
8	IVC R_x,R_y	ICo	Inverse conditionally the selected segment if encoding condition satisfies (ON bits >Seg. Size/2). R_x=NOI & R_y=Flags (R_x/R_y= R_0,R_1,...R_7).	IVC R0,R1
9	FL R_x,R_y	ICo	Flip selected segment bits. R_x=NOI & R_y=Flags (R_x/R_y= R_0,R_1,...R_7).	FL R0,R1
10	FLC R_x,R_y	ICo	Flip conditionally the selected segment bits if encoding condition satisfies (Seg. >Flip(Seg.)). R_x=NOI & R_y=Flags (R_x/R_y= R_0,R_1,...R_7).	FLC R0,R1
11	IVFL R_x,R_y	ICo	Invert and flip selected segment bits. R_x=NOI & R_y=Flags (R_x/R_y= R_0,R_1,...R_7).	IVFL R0,R1
12	SP H, EP, C/R_y	TH/WEEP	Send C or R_y number of pulses (R_y = R_0,R_1,...R_7, C = constant). Halt PC if H=1 (H=0,1). Send one extra pulse if EP=1 (EP=0,1).	SP 1,1,4
13	SD H, C/R_y	TH/WEEP	Inter-symbol separation of C or R_y number of clock cycles (R_y = R_0,R_1,...R_7, C = constant). Halt PC if H=1 (H=0,1).	SD 1,4
14	WR R_d, C	–	Write constant value to a register R_d (R_d= R_0,R_1,...R_7).	WR R0,8
15	SRD C/R_y	TH/WEEP	Set receiver inter-symbol separation equal to C or R_y number of clock cycles (R_y = R_0,R_1,...R_7, C = constant).	SRD R0 or SRD 4
16	WRI WE, EP, R_y	TH/WEEP	Wait for receiver pulse stream interrupt. PC halts till the interrupt arrives. Remove one extra pulse count if EP=1 (EP=0,1). Enable received pulse count write to register R_y (R_y= R_0,R_1,...R_7) if WE=1 (WE=0,1).	WRI 1,1,R0

(continued)

Table 8.2 (continued)

	Instruction	CB	Description	Example
17	SDB C	TH/WEEP	Sets the index bits or the data bits in the LoadReg as per the received pulse stream. (C=0,1 for indexing and data, respectively)	SDB 1
18	BNZD R_d, label	–	Branch to label and decrement R_d by 1 if the specified register R_d contains non-zero number. (R_d= $R_0,R_1,\ldots R_7$)	BNZD R0,loop
19	CRC R_x,R_y,I,Co	ICo	Copy register conditionally. R_x= R_y if I=0. R_x= R_y, if I=1 and LoadReg [R_y]=1 and Co=0. R_x=0 otherwise. R_x=Selected Segment, if Co=1. R_y is ignored (R_x/R_y= $R_0,R_1,\ldots R_7$). Can be used to clear the register.	CRC R1,R2,1,1
20	CF R_z,R_x,R_y	ICo	Combine Flags. R_z={$R_x[1:0]$, $R_y[1:0]$}. R_z=$R_0,\ldots R_3$. R_x/R_y=$R_0,\ldots R_7$.	CF R0,R1,R2
21	SF R_z,R_x,R_y	ICo	Split Flags. R_x=$R_z[3:2]$, R_y=$R_z[1:0]$. R_z=$R_0,\ldots R_3$. R_x,R_y=$R_0,\ldots R_7$.	SF R1,R2,R0
22	NOP	–	No operation.	NOP

Table 8.3 ECSIA interpretations

Instruction interpretation	
Control bit	Value: effect
T (Type R/C)	0: register, 1: constant
H (Halt-PC)	0: no halt, 1: halt
WE (write enable)	0: register write disabled, 1: enabled
EP (extra pulse)	0: extra pulse disabled, 1: enabled
I	0: no indexing, 1: indexing
Co	0: copy segment disabled, 1: enabled

stream and/or an extra pulse should be removed from the received pulse stream. The last 8-bit-long fragment of instruction is used to indicate a register number or an immediate constant value.

Type II

The second type of instructions handles two or three operands simultaneously and is used in operations such as encoding (inversion and reversion with or without condition), combining and splitting encoding flags, and copying register contents or some other information to a specified register conditionally. These instructions use CB in the second part of the instruction format where CB represents ICo. The combinations of I and Co bits are used to select the source of information to be copied. The 3-bit R_x and R_y, and 2-bit R_z $Register$ fields are used to indicate one of the general purpose registers.

Type III

The third group of instructions, highlighted in dark gray, handles two operands at a time and is used in operations such as updating a register with a given constant value and jumping to a specified label in the code depending on the validity of a condition specified by a register. Instead of CB, these instructions use R_d in the second part of the instruction format. The 3-bit *Register* field is used to indicate one of the general purpose registers, and the 8-bit *Constant* field is used to provide either a constant value or a label in the code to jump to.

8.2.3   Addressing Modes

ECSIA does not need any data memory. Therefore, the operands of all the instructions in Table 8.2 are either included in the instruction itself or accessed directly through the registers. As a result, ECSIA employs only three addressing modes: immediate, register, and auto-decrement. In the immediate mode, the source is either a constant or a label, while the destination is one of the general purpose, special purpose, or program counter registers. In the register mode, the register contains the value of the operand. The auto-decrement mode is used only for a jump operation where the branch to a label is taken and a specified register decrements by one if the register contains a non-zero number.

8.2.4   External I/O and Interrupts

As highlighted earlier, the ECSIA processor can be used as a stand-alone, streaming, core processor or as a co-processor in an SoC. The ECSIA I/O ports and the interrupts are designed to accommodate both the stand-alone and SoC configurations, using a simplified interface with three external I/O ports. One of these ports is the 16-bit data I/O port that is used to read from and write back to the external environment. To transmit and receive the packets in the form of pulse streams, a 1-bit *signal* I/O port is used. Another 1-bit *data-ready* port is used to source the generation of I/O interrupts and start the execution of instructions.

To interact with the I/O interface and achieve low-power operation, the ECSIA processor supports a workload-based interrupt mechanism consisting of three interrupt signals. The first is the I/O interrupt, which is generated when the data at the I/O port is available. The processor remains in a halt state until the I/O interrupt is received and starts instruction execution from the very start. The second is the transmitter interrupt, which is used to indicate the completion of the transmission of one pulse stream. If halt is enabled, the ECSIA processor remains in a halt state until the transmitter interrupt is received, at which time the execution is continued from the point it was halted. The third interrupt is the receiver interrupt, which is generated when the reception of one pulse stream is completed. The ECSIA processor remains in a halt state until the receiver interrupt is received, at which time the execution is continued.

8.2.5 ISA Discussion

ECSIA has a very compact RISC-like instruction set of only 22 instructions that can be functionally put in 4 different categories: (1) configuration: 8 instructions, (2) encoding/decoding: 8 instructions, (3) transmission control: 4 instructions, (4) register/branching: 2 instructions.

It is important to note that this DSA does not contain arithmetic/logic or rotation instructions. Nor does it have a call stack. Its branching instructions are limited to just one. This compact, minimalist approach to the ISA is a hallmark of the current DSA trend. Indeed, the TPU instruction set itself has less than 15 instructions [28] with the interesting twist that they are CISC rather than RISC instructions. The main reason for using CISC-style instructions in the TPU is to enable efficient instruction dispatching from the host memory to the TPU using a PICe bus. In our implementation, the ECSIA processor has its own instruction memory and program control.

In terms of execution latency, most of the ECSIA instructions execute in 1 clock cycle, which obviates the need for pipelining to improve data rate. In contrast, and as expected, the average number of execution clock cycles of the TPU CISC-style instructions is between 10 and 20 clock cycles. The TPU has a four-stage execution pipeline [29].

8.3 ECSIA Micro-Architecture

The reader is referred to Fig. 8.1 for the main blocks of the ECSIA processor. Figure 8.3 gives a more detailed breakdown of these blocks along with their connectivity, their access to signal and data ports, and their synchronous behavior. This micro-architecture further shows the clock distribution and program counter (PC) control units, the instruction decoder, the encoding/selection unit, the encoding/selection control, and the interrupt handler. The proposed micro-architecture executes all the instructions listed in Table 8.2. A few of these instructions use only one processing block at a time, while others use multiple blocks simultaneously to generate their outputs. As mentioned in the previous section, most of the instructions are executed in one clock cycle only. Instructions such as SP and SD take more than one clock cycle to execute as they need to send or receive a number of pulses. All the micro-architecture blocks of Fig. 8.3 are explained in the next subsections.

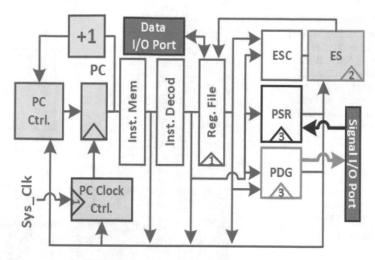

Fig. 8.3 ECSIA micro-architecture block diagram

8.3.1 Memory Interface

ECSIA supports up to 128 KB of instruction memory in a configuration of $64Kb \times 16$ with 16-bit address bus. ECSIA is a data-streaming processor. Only the instruction memory is needed, and there is no need for data memory. The program counter and the relevant controls are used to update the instruction memory address at each clock cycle.

8.3.2 Instruction Decoder

The instruction decoder accepts instructions from the instruction memory and decodes them to generate the appropriate control signals. The control signals are then used by several system blocks to take decisions and perform the required tasks. The ECSIA decoder, shown in Fig. 8.4a, processes the 5-bit opcode field in the 16-bit instruction and generates 27 control signals. The decoder logic works as per Table 8.4 where the opcode for each of the instructions is also listed. In Table 8.4, RR is *reverse the roles*, WB is *write back*, WE is *write enable*, RE is *read enable*, SE is *store enable*, LE is *load enable*, and SGE is *stage enable*. The remaining control signals represent a specific instruction. RE and WE enable the read and write operations of a register in the register file. WB enables the write-back operation to update the load register with the results after finishing a particular task. RR is used to reverse the roles of the registers in Type I and Type III instructions. In the CF, SF, and CRC instructions, the register locations can either be used to read from, or write to, a register. For example in CF, the second operand is used to read

Fig. 8.4 ECSIA micro-architecture hardware blocks: (**a**) Instruction decoder, (**b**) Register file, (**c**) Clock distribution and PC control, (**d**) Encoder and selector control, (**e**) Pulses and delay generator, (**f**) Pulse stream receiver

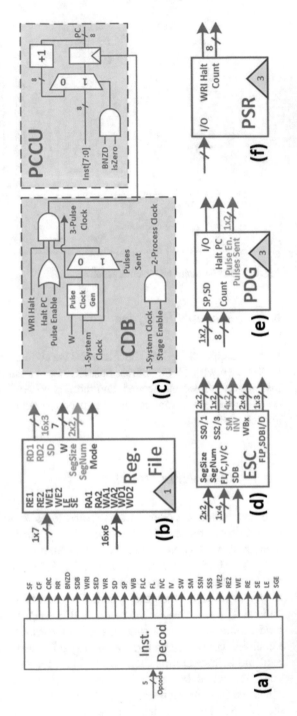

Table 8.4 ECSIA instructions decoder

Instr.	Opcode	Control signals[a]
NOP	00000	0000000000000000000000000000
RP	00001	0000000000000000000000000010
WP	00010	0000000000000000000000000100
SSS	00011	0000000000000000010000000000
SSN	00100	0000000000000000100000000000
SM	00101	0000000000000001000000000000
SW	00110	0000000000000010000000000000
IV	00111	0000000010010001000010100011
IVC	01000	0000000010010010000010100011
FL	01001	0000000010010100000010100011
FLC	01010	0000000010011000000010100011
IVFL	01011	0000000010010101000010100011
SP	01100	0000000000100000000000001000
SD	01101	0000000001000000000000001000
WR	01110	0000000010000000000000010000
SRD	01111	0000000100000000000000000000
WRI	10000	0000001000000000000000000000
SDB	10001	0000010000010000000000000011
BNZD	10010	0001100010000000000000011000
CRC	10011	0011000010000000000000011000
CF	10100	0101000010000000000000111000
SF	10101	1001000010000000000001011000

[a] [SF,CF,CRC,RR,BNZD,SDB,WRI,SRD,WR,SD,SP,WB,FLC, FL,IVC,IV,SW,SM,SSN,SSS,WE2,RE2,WE,RE,SE,LE,SGE]

a register, but in SF the second operand is used to store a part of the result. SE and LE are used to read and write the *LoadReg* in the register file. The SGE control signal is a power management signal that is used to shut-off unused modules of the architecture to save power during the execution of an instruction.

8.3.3 Register File

The ECSIA register file is shown in Fig. 8.4b and consists of all the registers listed in Table 8.1. The ECSIA register file supports read and write operations of two registers simultaneously and, therefore, has separate module ports for these. However, the *LoadReg* does not use any of these ports and has separate control signals. The output of the register file is divided into two parts. In the first part, the contents of the registers such as *LoadReg* and the general purpose registers are set to the output when requested through suitable control signals. In the second part, the contents of the special purpose registers are continuously reflected at the output of the register file so as to guarantee the availability of the communication parameters for system use without the need of issuing an extra request. These parameters include segment size, selected segment number, mode of transmission, and pulse width. We will later show how these parameters help in selecting and updating a specific portion of the data word without additional read/write cycles and delays.

8.3.4   Clock Distribution and PC Control

This module can be divided into two parts: a clock distribution block (CDB) and a PC control unit (PCCU), as shown in Fig. 8.4c. The clock distribution block, CDB, takes care of all the clocks inside the ECSIA system. CDB generates two types of sub-clocks using the main *System Clock* that is represented by "1" in the figure. The *Process Clock* is the gated system clock, and all the blocks execute using this. The process clock is denoted by "2" in the figure, and the gating is enabled if the "Stage Enable" signal is inactive. The second sub-clock is the *Pulse Clock* and is denoted by "3" in the figure. The pulse clock is used to generate pulse streams and inter-symbol separators. The pulse clock is the output of a pulse clock generator that takes a pulse width parameter as an input from the register file and generates a clock signal accordingly. Besides clock distribution, the CDB block is also responsible for controlling the halt state activation and deactivation as may be requested. Recall that the halt state is kept active in the reception mode but is activated in the transmission mode only if requested via either the SP or SD instruction.

The PC control unit, PCCU, takes care of updating the PC register. In an upcoming clock cycle, PC can either be incremented or updated by a jump address if the jump instruction $BNZD$ is executed. In case of reset or end of instructions (e.g., the transaction is completed), the control keeps the system in halt state unless an interrupt for a new transmission is received at which time the execution restarts from the very first instruction in memory.

8.3.5   Encoder and Select Control (ESC)

The encoder and select control (ESC) block is another specialized decoder that helps in generating the control signals for the most complex unit in the ECSIA processor, namely, the *Encoder and Selector (ES)*. The ES block itself is described in the next subsection. As for the ESC block, its I/O is shown in Fig. 8.4d, with the outputs being control signals for the ES block as per Table 8.5. The ESC inputs come from two sources. The segment size and segment number parameters come from the register file, while the encoding-specific control signals come from the instruction decoder. The generated control signals enable the ES block to identify the requested operation as well as the segment and segment size to which the requested operation should be applied.

Table 8.5 Decoder: encoder and select control

Inputs			Outputs								
Seg. size	Seg. num	FLP	SS0	SS1	SS2	SS3	SM	WB0	WB1	WB2	WB3
3-bit	3-bit	1-bit	2-bit	2-bit	1-bit	1-bit	4-bit	2-bit	2-bit	2-bit	2-bit
0	0	X	0	2	1	1	1	0	3	3	3
	1	X	1	2	1	1	1	3	0	3	3
	2	X	2	2	1	1	1	3	3	0	3
	3	X	3	2	1	1	1	3	3	3	0
1	0	0	0	0	1	1	3	0	1	3	3
		1	0	0	1	1	3	1	0	3	3
	1	0	2	1	1	1	3	3	3	0	1
		1	2	1	1	1	3	3	3	2	0
2	X	0	0	0	0	0	F	0	1	1	2
		1	0	0	0	0	F	2	2	2	0

INV = SM & (IV | IVC),   FLP = FL | FLV

8.3.6 Encoder and Selector (ES)

The encoder and selector (ES), shown in Fig. 8.5, is the core block of ECSIA. The ES module selects and processes a particular segment in a given data word as per the size and number specified by the user. It is used for the whole spectrum range of operations of bit processing and packet encapsulation, including:

1. Conditional and unconditional encodings
2. Generation of packet flags and the number of ON bit locations
3. Extraction of the index numbers of ON bits
4. Extraction of the decimal number of a segment
5. Conditional and unconditional decodings
6. Retrieval of packet flags and the number of ON bit locations
7. Bit setting according to the index number
8. Segment setting according to the decimal number
9. Conditional register copying

Operations (1) through (4) are typically executed at transmission, while operations (5) through (8) are executed at reception. As such, the ES block is the heart of the ECSIA core in the same way the ALU is the heart of a CPU core. Given that the ECSIA is a communication core, one of the important design decisions we have to make is whether the ES block should be duplicated for the transmitter and the receiver. To reduce HW resources and power consumption, the ES block is optimized such that it is used for both transmission and reception without any change whatsoever. ECS transmission and reception are designed so that they use the very same modules, including encoding/decoding and segmentation/combination. The main difference between the transmission and reception paths lies in the sources and destinations of the data. Accordingly, the ECS communication process has been

Fig. 8.5 ECSIA
micro-architecture hardware
blocks: encoder and selector

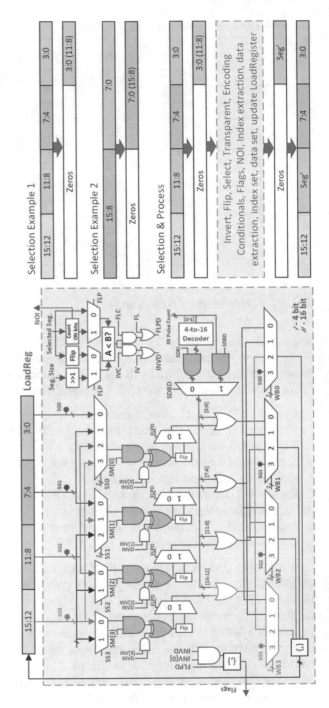

divided into three main phases, each with an implementation that is common to transmitter and receiver. The three phases are:

1. *Data composition*: The source is the I/O port during transmission and the pulse stream receiver (PSR) during the reception. The destination in both cases is *LoadReg* where data is composed for further encoding or decoding process.
2. *Encoding/Decoding*: During the encoding (transmitter) or decoding (receiver), both the source and destination are the same, namely, *LoadReg*. Segments are picked up from *LoadReg* and are replaced with the updated segments at the end of each encoding/decoding step.
3. *Output*: The source of data for both the transmission and reception is *LoadReg*. The destination is the pulse stream generator for transmission and the I/O port for the reception.

The block selects a segment from the *Load Reg* as per the set segment number and the segment size, relocates it to the LSB end, processes it as per the issued instruction, moves the updated segment back to its original position, and replaces the segment in the *Load Reg* with the processed one. Such a selection and place-back operation is shown in Fig. 8.5 (right), where in the first example, "Selection Example 1," where the segment size is 4 bits, the second segment is selected and moved to the LSB end with all other bits set to zero. A similar case is shown in the second example, "Selection Example 2," where the segment size of 8 bits and the first segment is selected for transfer to the LSB part. The bottom part of Fig. 8.5 (right), "Selection & Process," shows the process flow when applied to the first example.

ES Segment Processing

The selection and relocation of a segment to the LSB end at the start of the process and the place-back operation to the original location at the end are performed using a set of multiplexers shown in Fig. 8.5 (left). To perform this task, the control signals *SSx* from ESC are used for segment selection, and *WBx* are used for segment place-back. The second layer of AND gates uses the *SM* control signals to zero out all other bits except the selected segment so as to prevent any corruption of the results. The gray-shaded gates form a set of four identical logic functions that are used to AND/XOR all 4 bits of a data segment with the 1-bit control signal. The third layer of AND and XOR gates is used for bit inversion if requested and is controlled by the control signals *INV* and *INVD*. The role of *INVD* is to identify whether the requested inversion is conditional or unconditional. The fourth layer is comprised of "Flip" and MUX blocks and is used to perform the segment-wise flipping operation if requested and is controlled by the control signal *FLPD*. The same *FLPD* is used to decide if the flipping is conditional or unconditional. The *FLPD* and *INVD* are generated using circuitry that is shown at the top right corner of Fig. 8.5 (left). If the operation is to flip, the selected segment and its flipped version are compared to generate *FLPD* accordingly. If the operation is to invert, the number of ON bits in the segment is compared with half of the segment size and *INVD* is generated accordingly. As its name indicates, the *Count-ON-Bits* block counts the ON bits in the selected segment, and the shift-right block divides the segment size by two. The flags are generated by very simple AND and concatenate operations as shown at the left bottom side in Fig. 8.5. The inversion and flip operations are used in segment encoding and decoding.

ES in Receiver Mode

The process of reception is executed in two steps. In the first step, the received data is stored appropriately in *Load Reg*. To perform this step, the fifth layer in the ES block, which is comprised of a decoder, AND gates, OR gates, and a MUX, is used to either set a segment bit at index number specified by the received *RX Pulse Count*, or to translate this count into a decimal number and set the segment bits accordingly. For decimal number extraction, *SDBD* is used to pass the count through MUX, which is ORed with the selected segment bits. For setting a bit at an index, both *SDBI* and *SDBD* are used to pass, through MUX, only the ON bit that is generated by the decoder, which is then ORed with the selected segment. Recall that *Load Reg* gets cleared when the reception mode is selected and, therefore, the decimal number or the decoded bit at the output of MUX is ORed with zeros of the selected segment. This guarantees that the received data is not garbled and is stored successfully. During the first step of the reception mode, the middle layers of encoding go transparent and do not affect the selected segments. All the segments and their bits are updated iteratively in *Load Reg*. In the second step of the reception mode, the updated *Load Reg* is used to decode as was mentioned in the previous paragraph.

ES in Transmitter Mode

In this mode, the extraction of the index numbers of ON bits or the segment decimal numbers from the data loaded in *Load Reg* is performed using the instruction *CRC* whose functionality is described in Table 8.2. In executing *CRC*, the middle encoding layers and the received data-extraction layer go transparent. The selected segment is directly copied as a decimal number to a specified register that could be used later to transmit the pulses. To extract the index numbers of the ON bits, a program loop is used in which for each iteration, a bit in the selected segment at the index specified by the loop index number is checked. If the bit is ON, the iteration number is copied to the specified register. Otherwise, 0 is copied. This register could be used later to transmit the pulses. It must be noted that the place-back operation of *Load Reg* gets disabled during the extraction operations in the transmitter mode so as to protect the data. With such transmission mode, the ECSIA processor can transmit decimal numbers, ON bit index numbers, or a combination of both to form a legal packet according to the specification of the pulsing-index protocol.

8.3.7 Pulse and Delay Generator (PDG)

The pulse and delay generator (PDG) block transmits a pulse stream or an inter-symbol separator. The module interface diagram is shown in Fig. 8.4e. When SP or SD control signal at the input is set active by the instruction decoder, the module collects the count information from its input port and generates a pulse stream that is comprised of a number of pulses equal to the count. This count at the input is set by the instruction SP either through a provided immediate constant or by reading a specified register content, as described in the instruction set section. The pulse clock generated by the clock distribution block is used to transmit the pulse stream and hence is routed to the PDG block through CDB. The width of the pulses is the same as the pulse clock and, hence, is already set by the clock distribution block. Toward this end, the output signal *Pulses Sent* of PDG helps CDB to route the pulse clock when the SP instruction is issued and re-routes to system clock when the transmission ends. If specified in SP instruction, the signal *Halt PC* is used to halt the system until the transmission is complete. At the completion of transmission, instruction fetch is normally continued. The execution mechanism to send inter-symbol separator using SD is the same as that of SP except that the signal *Pulse Enable* is inactive. *Pulse Enable* gates the pulse clock for a number of pulses specified by the count, and as a result, a delay is generated at the I/O port. The I/O port of the block is directly connected to the output port of the ECSIA processor.

8.3.8 Pulse Stream Receiver (PSR)

The pulse stream receiver (PSR) block receives one pulse stream at a time. The block interface diagram is shown in Fig. 8.4f. The ECSIA input is connected to the I/O port of PSR. To receive a pulse stream, the edges of the incoming pulses are counted that end with the detection of an inter-symbol separator as described in the review section. To detect such a delay, an inter-symbol separator duration in terms of a number of pulses is necessary to know. This separator duration information is set by the execution of SRD instruction that updates the internal delay comparison register *RxDelay* of PSR. The WRI instruction, wait for receiver interrupt, activates this block and puts the system in halt state through CDB using the block's output signal *WRI Halt*. PSR keeps counting the stream pulses until the separator is detected. The separator is detected by counting the edges of local pulse clock when the input signal is low and comparing the count continuously with the RxDelay register. At the detection of a separator, PSR brings the system back to the normal state by generating WRI interrupt and updates the output port *Count* of the PSR block with the count of stream pulses. This count is then stored in the specified register if the option to write enable (WE) is set in the instruction. Otherwise, the count is discarded. The WE helps in receiving all the pulse streams in a particular communication packet but storing the count of desired streams only. The examples of such pulse streams include the start, stop, and synchronization patterns.

8.3.9   Interrupt Handler

The three types of interrupts (I/O, transmitter, and receiver) and their operations are described earlier in the ISA section. These three interrupts are taken care of using a small finite state machine that in collaboration with CDB, PSR, and PGD blocks helps to halt and resume the system operation. The ECSIA interrupt handler is very simple and does not use any interrupt vector or priority queue since only one interrupt can occur at a time, and such an interrupt can be handled using elementary logic.

8.3.10   Micro-Architecture Discussion

As pointed out in the introduction, the major benefit of a domain-specific instruction set is that it provides hardware designers with additional opportunities for optimizing the micro-architecture to maximize performance. In the case of ECSIA, this optimization is clear in the specialized blocks for encoding and selection (ES), pulse generation (PDR), and pulse reception (PSR). The ES block is the core of the ECSIA micro-architecture as it is the one responsible for all processing needed to either encode or decode packets. All the instructions from the encoding/decoding category are handled through the ES block. Similarly, the PDR and PSR blocks are specialized hardware modules that attend to the fundamental nature of the communication protocols under consideration, namely, the sending and receiving of pulse streams. They can be considered hardware shortcuts for the execution of instructions belonging to the transmission control and configuration categories.

In a similar vein, the *Matrix Multiply Unit* is the heart of Google's TPU accelerator and is the main execution block of the *MatrixMultiply* domain-specific CISC instruction [29]. As mentioned earlier, the TPU acts as a co-processor and has no instruction memory of its own. Rather the TPU CISC instructions are dispatched by the host processor to the TPU through the PCIe bus. On the other hand, ECSIA is implemented as a full, single-cycle, streaming processor with its own instruction storage, fetching, and decoding. Although larger word lengths can conceivably be chosen for the ECSIA data bus, the 16-bit word embodies domain knowledge related to the maximization of the transmission data rate of the edge-coded signaling protocols. The reader is referred to Chap. 2 for the algorithmic and experimental details on ECS data rate maximization.

The I/O ports of the ECSIA micro-architecture that are explained in Sect. 8.2.4 and illustrated in Fig. 8.3 represent the minimal number required for a functional processor. There are a data port, a signaling port, and a received data-ready interrupt port. This interface is sufficient for both stand-alone operation, embedded operation, or SoC integration. However, other interface designs are conceivable, including the following:

1. Support for an external interrupt to start data transmission.
2. Support for data re-transmission without re-coding or re-segmentation.
3. Support for internal interrupt mechanisms for data transmission and reception.
4. Support of serial and/or parallel data port options along with a SerDes module implementation.
5. Support of re-segmentation and re-encoding of *LoadReg*.

Finally, note that the ECSIA opcode leaves room for ten more instructions, and of course, its ISA can be extended to cover other functionalities such as information security. A ECS cryptographic block [59] may be added to the micro-architecture. Such a block would be enabled and initialized with its own special instructions. A similar ISA extension approach has been recently used to secure the RISC-V processor [46].

8.4   Experimental Verification and Results

Verilog HDL is used to code a fully functional processor based on the proposed DSA and ECSIA micro-architecture. A full testbed is set up using the Xilinx Spartan-6 FPGA platform. The prototype platform is used to verify ECSIA functionality and performance. Extensive simulations and real-time hardware verification are performed to confirm the results. A clock rate of 25 MHz is used for the ECSIA testbed. The verification methodology requires that the ECSIA transmitter sends 16-bit data words starting at 0 with an increment of 1 at each transmission. The ECSIA receiver resends the same data back. The returned and original 16-bit data words are compared to check for any bit errors during the round-trip.

In another experiment, the software aspects of the two implementations are compared. In one implementation, the ECS protocols are programmed on a low-power RISC processor, namely, TI's MSP432X. In another implementation, the same ECS protocols are programmed using the ECSIA assembly language and run on the ECSIA processor. Both implementations use a 25 MHz clock. The number of instructions required to implement these techniques using MSP432X is approximately 1100 to 1400 on average, whereas ECSIA needs only 50 to 100 instructions. The reduction is by a factor of 13 to 28. The data rates of the MSP432X implementation have suffered as a result of clock cycles being "wasted" on computation rather than used for communication. The data rate reduction was by a factor of 100. On the other hand, the ECSIA data rates are maintained close to a hardwired implementation of the ECS protocols. This is a direct result of the ECSIA optimized hardware and in line with the DSA philosophy of having the processor to do one task extremely well. The software implementation comparison is shown in Table 8.6 and Fig. 8.6.

Table 8.6 Results

Software implementation comparison			
	ECSIA	MSP432X	% Change
No. of instructions[a]	50–100	1100–1400	92.8–95.4[b]
Data rate (Mbps)[a]	≈4.1–7.1	≈0.041–0.071	99[c]
Hardware synthesis comparison			
	ECSIA	Stand-alone	% Change
Power (μW)	≈31.14	≈19–26.6	14.6–38.9[c]
Avg. E_b (pJ/bit)	≈4.3–7.6	≈2.7–6.5	14.5–37.2[c]
Area (gate count)	≈4700	≈2100–2400	48.9–55.3[c]
	ECSIA	Combined	
Power (μW)	≈31.14	≈30[d]	3.6[c]
Avg. E_b (pJ/bit)	≈4.3–7.6	≈4.2–7.3	2.3–3.9[c]
Area (gate count)	≈4700	≈6600	28.8[b]

[a] Average
[b] Increase
[c] Decrease
[d] Control logic + One active ECS protocol

Fig. 8.6 ECS family implementation: ECSIA vs. MSP432X

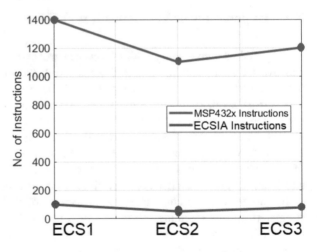

Further, the ECSIA processor has been synthesized using GLOBAL-FOUNDRIES 65 nm technology and estimated to consume around 31.14 μW with a gate count of about 4700 gates. The power estimate has been obtained from the Synopsys design compiler power report. This estimate includes both dynamic power and leakage power. Furthermore, the dynamic power includes both standard-cell switching power and interconnect switching power. A hardware solution combining the three edge-coded family members (ECS1, ECS2, and ECS3) has approximately 6600, which exceeds the ECSIA core gate count by more than 40%. Given that two of the three hardwired protocols are gated on the combined chip to save power, the ECSIA core will be comparable in terms of power and ahead in terms of area. The hardware synthesis comparisons are shown in Table 8.6. Note that the core logic voltage of the ECS DSA is 1.1V, which is the default VDD of GLOBALFOUNDARIES 65 nm CMOS technology. The I/O voltage of the ECSIA processor is 3.3 V, which, in our experiments, is the I/O voltage value provided on the Xilinx Spartan-6 FPGA board.

The ECSIA power consumption remains well within the power budget of a fully hardwired implementation of a stand-alone edge-coded signaling protocol. Other advantages of the DSA are reduced gate count, preserved data rate, and reduced programming effort. Shorter programs are less likely to have bugs than long programs, and as a result, the ECS development cycle using ECSIA is likely to be much shorter. The ECSIA solution offers a fully programmable communication interface that is specifically geared to the realization of pulsed-transmission techniques. Such capability can of course be used to implement not only ECS1, ECS2, and ECS3 but also any custom nonstandard protocol without the need for any change in hardware. Table 8.6 shows a comparison between the ECSIA codes of the ECS protocols and their full hardware counterparts.

The maximum clock frequency of the ECSIA processor depends on the clock frequency limits of the hardware implementing ECS. In our experiments, we have used the Spartan-6 FPGA kit that allows a maximum clock frequency of 100 MHz. However, migrating to a high-end FPGA such as Virtex-7, the maximum speed would increase to more than 500 MHz. Similarly, for an ASIC implementation, the underlying hardware technology defines the maximum clock frequency. In our experiments, we have synthesized the ECSIA processor for a maximum clock frequency of 100 MHz.

Processor performance is usually analyzed using the CPU performance equation for a given workload [23]

$$t = IC \times CPI \times T \tag{8.1}$$

where t is the workload run time on the processor, IC is the workload's instruction count, CPI is the average number of clock cycles per instruction, and T is the clock time period. Any of the three parameters can be tuned to improve CPU performance. In an IoT network, the majority of the devices use low-end processors with few tens of MHz of the clock frequency and, therefore, the parameter T cannot be decreased beyond a certain limit. On the other hand, the reduction in the total number of instructions IC can bring a remarkable increase in performance. This is one of the major benefits of a DSA. Under the assumption that the MSP432X processor has $CPI = 1$ and uses the same 25 MHz clock as ECSIA, then MSP432X would need 52–56 μs to execute the 1100–1400 program instructions, while the ECSIA needs only 2–4 μs to execute the same task using its 50–100 instructions. Since ECSIA is a communication core and the ECS data rate is dynamic, the more appropriate performance comparison is to use the following equation:

$$t_{WL} = C \times WorkLoad \times T_P \tag{8.2}$$

Table 8.7 Time to finish
workload

	General purpose	ECSIA DSA	Speedup
ECS1	≈164 s	0.5243 s	312.8
ECS2	≈ 90 s	0.2862 s	314.5
ECS3	≈103 s	0.3277 s	314.3

where t_{WL} is the run time of the workload, C is the average number of clock cycles needed to transmit 16-bit data words, T_P is the time period of these pulses, and $WorkLoad$ is $2 \times 2^{16} = 131072$ 16-bit data words sent on a round-trip along the ECS communication link. The C for ECS1, ECS2, and ECS3 are 100, 55, and 63, respectively. The T_P for MSP432X and ECSIA are 12.5 μs and 40 ns, respectively. This is because MSP432X needs more instructions than ECSIA to generate ECS pulses, while the ECSIA micro-architecture has a specialized block dedicated to pulse generation that is activated with a single instruction. The t_{WL} results are shown in Table 8.7. It is clear that the proposed DSA for edge-coded signaling outperforms the general purpose CPU while remaining within the low-power budget of a hardwired protocol. As the testing workload is made of all possible combinations of 16-bit words, all other workloads will be a subset of the one we have used. Therefore, we estimate that the speedup numbers of Table 8.7 will remain qualitatively the same under any testing workload.

8.5 Conclusions

The Edge-Coded Signaling Interface Architecture (ECSIA) is a domain-specific architecture for single-channel, low-power, high data rate, dynamic, and robust communication based on edge-coded signaling protocols. Its RISC-style ISA is designed to facilitate the efficient coding of user-defined programs that are specific to such communication interfaces. The ECSIA micro-architecture is comprised of optimized processing blocks in support of the economical execution of such programs. The ECSIA supports both standard and customized edge-coded signaling protocols and enables the amalgamation of software and hardware to significantly reduce the number of instructions required to execute a given task while preserving the data rate and reliability of a bare-silicon design. The ECSIA processor has been synthesized in GLOBALFUONDRIES 65 nm technology and has been found to consume only 31.14 μW that translates into less than 10 pJ per transmitted bit. The ECSIA micro-architecture has been evaluated and compared with bare-silicon designs using a domain-specific metric based on the communication workload. The ISA can be extended to include support for a cryptographic block [59] that can be readily added to the ECSIA micro-architecture. One important direction for future research is to explore the usage of ECSIA as a lightweight, communication core either in an IoT communication hub or in a heterogeneous computing and communication environment.

Chapter 9
Application: Hardware Platform for IoT Sensor Networks

Pervasive is persuasive.

David Rose

The goal of this chapter is to present an FPGA hardware platform for the prototyping and analysis of ultra-low-power IoT sensor networks [50]. The platform, named Prototyped IoT (PIoT), is composed of ultra-low-power MSP430 microcontroller cores, acting as sensor nodes, and a single-wire communication protocol from the ECS family. The proposed platform is scalable in that the resources needed to implement the protocol for each node are minimal. Furthermore, with sufficient resources, it enables the analysis of the performance, timing, and power of IoT sensor networks that can have up to hundreds of nodes. The platform is equipped with a Multi-Core Debug Control Unit (MCDCU) to support the parallel debug of multiple sensor nodes. The hardware realization of PIoT is based on ECS1. However, the methods and results are valid across the entire ECS family.

9.1 Platform Architecture

The proposed platform (PIoT) is composed of eight FPGA based cores of MSP430, a low-power microcontroller family by Texas Instruments, connected to each other using an ECS link [57]. The platform is shown in Fig. 9.1. Each of the microcontroller cores represents a sensor node. An arbitrary number of these cores can, in principle, be instantiated without the need of any additional hardware circuitry. The number of instantiated sensor nodes is limited by the application requirements and the hardware resources available on the selected FPGA device.

Interestingly, the network of MSP430 cores does not require a separate connectivity configuration for each of the available topologies (i.e., master–slave, tree, point to point, etc.), and the configuration shown in the Fig. 9.1 can be used without any change. However, to support a particular topology, the embedded software may need to have some special checks and balances. Each of the nodes in the network is

© Springer Nature Switzerland AG 2022
S. Muzaffar, I. M. Elfadel, *Secure, Low-Power IoT Communication Using Edge-Coded Signaling*, https://doi.org/10.1007/978-3-030-95914-2_9

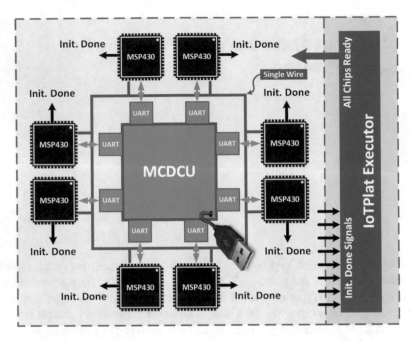

Fig. 9.1 Platform architecture

assigned a unique ID that is used for its identification across all communication and control tasks.

The UART (RS-232) protocol is used to program and debug the embedded software executed on the MSP430 cores. The FPGA boards, Xilinx Virtex-7 VC707 for example, are usually equipped with only one UART port for serial port communication. This limits the number of connections that can be established with the computer software tools. A debug module, known as the Multi-Core Debug Control Unit (MCDCU), is responsible for properly assigning the UART connections to the cores to resolve the debug connectivity issues. The MCDCU, shown in Fig. 9.1, uses an on-board USB-to-UART bridge to connect to the computer via a USB cable. Two different methodologies for MCDCU are adopted, which are discussed in Sect. 9.2.2.

A PIoT Initiator is used to put all the nodes in a wait state until all the nodes are programmed using the aforementioned program and debug interface. The module is shown in Fig. 9.1. Nodes are programmed step by step and executed immediately. Therefore, the embedded software of each node should wait for an *All Chips Ready* signal from the initiator, announcing that all the chips are ready to run. The *All Chips Ready* signal is generated based on an *Init. Done* initialization ready signal from each of the sensor node cores.

9.2   Platform Implementation and Testing

The PIoT is implemented using the Xilinx Virtex-7 VC707 evaluation board. Verilog implemented core of MSP430, OpenMSP430 [60], is used as a sensor node. The MCDCU, which provides the debug connectivity to all the instantiated cores, is implemented in two different ways: first, through an on chip connectivity manager, and second, through an external debug daughter card. Both techniques are discussed in the next subsections.

9.2.1   Sensor Nodes

Microcontroller Configuration

OpenMSP430 provides a Verilog HDL core and can be configured to any member of the MSP430 microcontroller series. The configuration is usually performed by changing the settings of the parameters available in the define files provided with the HDL core package. For the PIoT system, the MSP430F2330 microcontroller is configured. However, the core's peripheral wrapper is customized according to the needs of the system to support the MCDCU connectivity. OpenMSP430 supports I2C or UART as a program and debug interface. UART is selected as the default debug interface since I2C requires a special external I2C debug probe.

I/O Ports Selection

MSP430F2330 has only four 8-bit general purpose I/O ports some of which will be required for the external logic needed to control the nodes. The I/O ports can be reconfigured as per application requirements. P1[0] (Single Wire) is the only port dedicated to the communication between all the sensor nodes and an embedded C implementation of ECS1. To let the PIoT Initiator know that the node is ready to run, port P3[0] (Proc. Init. Done signal) is set to 1 when initialization is complete. Depending on the status pf *Init. Done* signals of the other nodes, the PIoT Initiator sets the P2[0] (*All Chips Ready* signal) to instruct all the cores to start operation. *All Chips Ready* can also be configured with on-board switches to bypass the PIoT initiator and control the operation manually.

Microcontroller Memories

Each MSP430 core needs two kinds of memory: program memory and data memory. Both are created using the Xilinx IP Core generator and connected internally with the MSP430. The program and data memories are of size 8 and 1 KB, respectively. Memory sizes are determined by the specifications, provided by the

Texas Instruments, for the MSP430F2330. Memory sizes are configured using the same define files used to configure the microcontroller architecture.

9.2.2   Multi-Core Debug Control Unit (MCDCU)

Sensor node cores can be programmed and debugged through the available on-board UART port, which is composed of a USB-UART bridge that can connect to and program only one core at a time. To enable the PIoT platform, multiple USB-UART connectors are needed, which can be achieved according to one of the following two methodologies.

On-Chip MCDCU

A UART port multiplexer for all the TX signals and a de-multiplexer for all the RX signals are implemented to support the programming of all the available sensor nodes. The multiplexer and de-multiplexer selection pins are connected to the on-board switches for the selection of a particular sensor node. Depending on the configuration of switches, the UART port is connected to the selected node, and the node can be programmed or debugged. The limitation associated with this method is that only one core can be debugged at a time. Also to program all the cores, one needs to change the configuration of switches for each of the cores. To remove this limitation, an external debug daughter card is developed, and this is described in the next subsection.

External MCDCU Daughter Card

The external MCDCU daughter card allows to connect, program, and debug multiple cores simultaneously, as shown in Fig. 9.2. The MCDCU daughter card is composed of eight USB-to-UART bridge ICs (CP2102 from Silicon Labs), Mini-USB type B connectors, and I/O headers. The I/O headers on the daughter card connect to the general purpose I/O headers on the FPGA board to provide UART (RX and TX) connections. On the other end, the daughter card USB ports connect to the computer via USB cables with each of the USB ports detected as an UART COM port and assigned with a unique COM port ID. Using these COM ports, the software tools can connect and communicate to the cores independently. The number of established debug connections is not limited to eight as more than one daughter cards can be used to provide more debug connections.

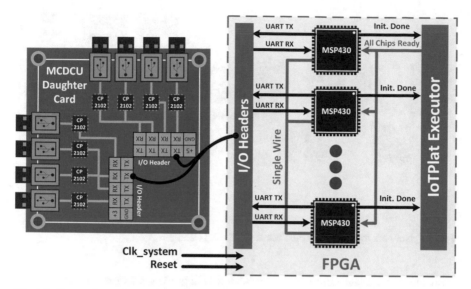

Fig. 9.2 System integration

9.2.3   Embedded C ECS1 Transceivers

The PIoT inter-nodes communication is carried out using the ECS1 transceivers, which are composed of a transmitter and a receiver and implemented in the embedded C language for this experiment. The software implementation of ECS1 transceivers allows to instantiate multiple sensor nodes without any external ECS1 physical layer. The transceiver provides two very simple function calls to engage the ECS1 transmission, *SingleWire_TX(tx_data)* and *SingleWire_RX()*. Reception is the default transceiver mode. An interrupt service routine is used to count the number of pulses of each input pulse stream, which are later decoded to infer the data word. The transmitter follows the same steps, as described in Chap. 2. Before the transmission is started, the *Single Wire* port is configured as output, and the interrupts are disabled to bring the ECS1 in transmitter mode.

9.2.4   System Integration

The complete PIoT is composed of an FPGA-based sensor network and an external MCDCU daughter card, as shown in Fig. 9.2. The UART signals of each of the nodes are connected to the FPGA board I/O headers, which further connect to the I/O headers on the MCDCU daughter card. For the proposed platform, eight sensor nodes are instantiated. To make these inserted cores work successfully, the UART, *Single Wire*, *Init. Done*, and *All Chips Ready* signals are necessary to connect

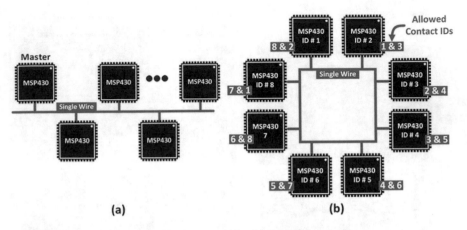

Fig. 9.3 Network topology : (**a**) PIoT master–slave (**b**) Configuring ring

with the appropriate modules. The MCDCU daughter card does not require any additional power supply and operates using the computer provided USB VBUS power source. To debug more than eight cores simultaneously, more daughter cards can be connected to the FPGA board using the I/O headers.

A master–slave network topology is configured using embedded C programs for the sensor nodes to test and analyze the platform. As shown in Fig. 9.3a, one sensor node acts as a master and sends the request on the ECS1 bus followed by a reply from the slave sensor node. To engage a particular sensor node, the master node includes its unique ID in the request packet. On the other hand, all the slave nodes receive, match their IDs, analyze the request, and send back the required information.

PIoT is not limited by the network topology used as any other topology can be configured without any change in the hardware architecture. As mentioned previously, this topology configuration is performed in the embedded C programs for the sensor nodes. Ring topology, for example, can be used by limiting each of the sensor nodes to use only two other node IDs for communication. Such configuration is shown in Fig. 9.3b, where each node is capable of contacting only two other nodes, using the same single-wire line. It must be mentioned here that message collision is prevented using the built-in properties of ECS1. Additionally, other collision prevention techniques can also be applied easily by updating the ECS1 embedded transceivers and the logical topology control (LTC) unit accordingly.

9.3   Compiler and Debugging Tools

The Eclipse IDE development environment is used to code, verify, and compile the MSP430 program. The MSPGCC compiler version 20120502 is configured with the eclipse IDE. Additionally, the MSP430 plug-in for the eclipse IDE [61] is required to generate the proper settings and paths for the eclipse IDE to recognize the compiler and the target microcontroller architecture. Once the software is compiled, an executable (.elf) file is generated and loaded into the program memory of the microcontroller with the help of the OpenMSP430 debugging tools [61].

Ubuntu Linux is used to run the TCL debugging scripts. The scripts are provided with the core package. The debugging tool, openmsp430-minidebug, is a minimalistic debugger with a simple GUI, used to select and load the eclipse generated executable (.elf) files. Minidebug loads the program into program memory and waits for the run command to start execution. The tool also provides a debug environment that is restricted to debugging the assembly code only.

9.4 Conclusions

PIoT is an FPGA platform to prototype, characterize, and evaluate the network design options for ultra-low-power IoT sensor networks using single-channel communication protocols. HDL cores of TI's MSP430 act as sensor nodes and communicate with each other using the Pulsed-Index Communication protocol. Embedded C implementation of PCS1 eliminates the need of any external physical layer and offers easy-to-use communication calls. A Multi-Core Debug Control Unit (MCDCU) provides the interface to connect, program, and debug multiple sensor nodes simultaneously. The platform is also scalable in that the resources used for a two-sensor, point-to-point communication link is less than 1% of the Virtex-7 available hardware, and therefore, a large number of sensor nodes can be added conveniently. PIoT facilitates network reconfiguration, which enables the analysis and design of a variety of network topologies, including buses, stars, rings, and trees. We are currently using the platform to analyze IoT networking options from the viewpoints of their data rates, power consumption, synchronization, and robustness to node failures.

Chapter 10
Application: Body-Coupled Communication

The medium is the message.

Marshall McLuhan

In this chapter, we show how edge-coded signaling can help in addressing the challenges of designing reliable, low-power transceivers for body-couple communication (BCC). Specifically,

1. We report on the very first BCC transceiver design based on the Edge-Coded Signaling (ECS) protocol.
2. We report on the very first experiments in round-trip, BCC transceiver testing using arbitrary bit patterns.
3. We report on a full BCC working prototype capable of transmitting over a range of 150 cm with zero bit-error rate.

Our simplified, low-power, and low-complexity BCC transceiver is self-synchronizing and does not require any circuitry for clock and data recovery or for duty cycle correction.

10.1 Introduction

Wearable devices have always been the focus of active research, and technology advances have made it possible to develop sophisticated wearable electronic devices such as smart watches, smart eyeglasses, and fitness/lifestyle monitors. The emergence of IoT has significantly enlarged the scope of research on wearable electronics to include a new range of smart wearables such as caps, clothes, shoes, headphones, ornaments, and healthcare sensors. Reliable real-time communication among these body-worn devices plays a key role in the synchronous collection of information about the human body and its environmental conditions, and therefore, in the enablement of a new era of portable diagnosis and personalized care. The ability to transmit and receive data at a very low energy-per-bit rating is an essential characteristic

© Springer Nature Switzerland AG 2022
S. Muzaffar, I. M. Elfadel, *Secure, Low-Power IoT Communication Using Edge-Coded Signaling*, https://doi.org/10.1007/978-3-030-95914-2_10

of such wearable devices as they need to remain operational during days, perhaps weeks, of continuous usage. To date, the state-of-the-art healthcare platforms use either a network of wires or wireless protocols to establish communication links between wearable devices. Existing wireless standards are power-hungry [41] and are known to drain the batteries quickly while wired communication is in conflict with the stringent wearability requirement. An alternative to wired or wireless communication is body-coupled communication (BCC), which uses the human skin as a communication medium.

Several techniques enabling BCC have already been proposed in the literature [25, 42]. The focus of these single-channel BCC techniques is on recovering the data bits once the bit stream is synchronized with the local receiver clock. A full BCC link has therefore to engage the use of complex and power-hungry circuitries for clock and data recovery (CDR) and duty cycle correction. Sophisticated modulation schemes such as OFDM [68, 69] and WDM [42] have been proposed for BCC to address issues such as multipath fading, variable-ground effect, and variable skin-electrode impedance, but the resulting transceivers are rather complex and very challenging to design as they require finely tuned, mixed-signal design methodologies and tools. Additionally, in the majority of the proposed BCC transceivers, the testing is performed either using a periodic clock waveform through the body or by performing spectral analysis. To the best of our knowledge, a full demonstration of a successful bi-directional transmission of data bits, involving an arbitrary, or pseudo random, number of ON and OFF bits has not been entirely achieved so far. We believe the main reason to be that the transceiver behavior varies with the variable intervals of ON and OFF signals. Moreover, while spectral analysis of BCC transceivers is important for verifying transceiver compatibility with the frequency-domain characterization of the body channel, it does not in itself guarantee successful time-domain operation. Therefore, there is a need for a simplified BCC transceiver, which can be used to establish an error-free, real-time, bi-directional communication link between body-worn devices while meeting stringent ultra-low power and energy efficiency requirements.

In this chapter, we address the above challenges and make the following novel contributions:

1. We report on the very first BCC transceiver based on the Edge-Coded Signaling (ECS) protocol.
2. We report on the very first experiments in round-trip, BCC transceiver testing using arbitrary bit patterns.
3. We report on a full BCC working prototype capable of transmitting over a range of 150cm with zero bit-error rate.

Our simplified, low-power, and low-complexity BCC transceiver is self-synchronizing and does not require any circuitry for clock and data recovery or for duty cycle correction, thus fully exploiting the salient features of ECS.

10.2 ECS Signaling and BCC

The characteristics of the human body channel is that of a band-pass filter with cut-off frequencies at 10 KHz and 120 MHz. Therefore, a digital bit stream passing through a human body channel is distorted, and the output signal is a train of positive and negative spikes aligned with the rising and falling edges of the digital input signal. The voltage magnitude of these spikes is typically in the neighborhood of ± 50 mV with a settling time of about 8 ns. This limits the maximum achievable date rate to about 125 Mbps. Received spikes that are within 8 ns of each other are indistinguishable. BCC is complicated by the fact that there is no common electrical ground between transmitter and receiver, and so the body acts as an antenna picking up 50 Hz and other AC noise signals, thus contaminating the BCC output.

The basic circuit to transmit and detect BCC signals is shown in Fig. 10.1 and is comprised of a front-end filter, an amplifier, and a Schmitt trigger for edge detection. It is important to note that such circuit needs no data conversion at the input or output as the data pulse stream is directly injected through the transmitter electrode at the input and recovered after edge detection through the receiver electrode at the output. The main complication in such a circuit is that the duty cycle of the recovered output is variable, and as a result, there are higher chances of bit errors when traditional modulation techniques are used. To mitigate such a negative impact of duty-cycle variabilities and reduce bit-error rate (BER), duty-cycle correction circuitries have been typically used with BCC transceivers. Moreover, the BCC receiver needs to synchronize the recovered signal with its local clock to successfully infer the data bits. This synchronization is typically accomplished using clock and data recovery (CDR) circuitries. Communication standards that do not use any CDR, such as UART and Dallas 1-Wire, depend on accurate duty cycles to time and infer the data bits. This dependence and the presence of a hard lower bound on the duty cycle itself significantly limit the data rates of the common single-channel standards. Complex modulation schemes such as OFDM and WDM that are meant to address

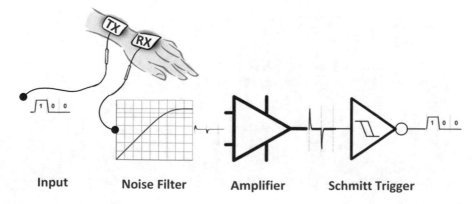

Input   Noise Filter   Amplifier   Schmitt Trigger

Fig. 10.1 Base architecture for recovering BCC signals

such issues as multipath fading and variable-ground electrode cannot do away with the CDR and duty-cycle correction circuitries, with the net result being a further increase in transceiver power consumption. Furthermore, the time-domain testing of existing BCC transceivers as reported in the literature has typically been done with periodic clock signals and rarely with pseudo-random data bit streams. In particular, the variation in the duration of the ON or OFF signal intervals has not been captured in testing signals. To our knowledge, none of the existing BCC systems has been tested for bi-directional signaling. Our proposed transceiver using the ECS signaling technique is tested under both requirements of a pseudo-random bit stream and a round-trip signaling through the human body. The hardware realization of the transceiver is based on the ECS1 protocol. However, the methods and results are valid for the entire ECS family.

10.3 BCC Transceiver

The most important point of this chapter is that a judicious selection of the bit encoding protocol results in a significant simplification of the BCC transceiver. In particular, the selection of the edge-based encoding protocol as implemented in the ECS family results in a streamlined analog transceiver architecture that does not require the data conversion, data synchronization, or duty cycle correction blocks. These blocks are typically found in the published BCC transceivers that use the well-known digital modulation/demodulation schemes. In the rest of this section, we will show how the optimization of the basic transceiver of Fig. 10.1, in combination with ECS1, results in a successful bi-directional BCC having single-digit mW of power consumption and sub-$1nJ/bit$ of energy efficiency.

The circuit diagram of the proposed BCC transceiver is shown in Fig. 10.2. A passive 10 KHz–100 MHz band-pass filter is used to filter out the 50 Hz and other AC noise signals from the received signal through the skin channel. Next, the

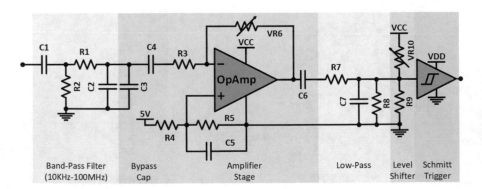

Fig. 10.2 BCC receiver circuit

filtered signal is amplified to achieve voltage levels within the hysteresis range of the Schmitt trigger, which is used to recover the pulses from the positive and negative spikes of the distorted signal. Typically, the amplifier is operated in saturation mode, but the feedback resistor of the amplifier stage is tunable and helps in adjusting the voltage levels as per the requirements. The voltage levels are adjusted such that the peaks of the positive spikes are above a maximum threshold, the valleys of the negative spikes are below minimum threshold, and the zero voltage remains in the middle of the Schmitt trigger's hysteresis range. This adjustment is facilitated using a voltage-level shifter that comprises two resistors, one of which is variable and is used to fine-tune the shift level. A low-pass filter is used between the output of the OpAmp and the input of the level shifter to mitigate the impact of amplifier-generated noise, which in our case is in the spectral range of 300–400 KHz but can vary from one amplifier to another. In some cases, the amplifier-generated noise is itself significantly amplified due to the saturation mode of operation and, therefore, needs cancellation. Otherwise, an intermediate filter after the amplifier is not required. The last stage is the Schmitt trigger that converts the spikes into digital bits of variable duty cycle. The variability of the duty cycle of each bit depends on the location of input spikes and the hysteresis range of Schmitt trigger. Note that the Schmitt trigger operates at a lower voltage supply than the Op Amp in order to shrink the hysteresis range and successfully accommodate the voltage levels of the input spikes. The main reason of this differential in supply voltages is that the amplitude of the input spikes decreases as they pass through the low-pass filter. This decrease cannot be compensated in the amplifier stage due to the cap in the amplifier gain. The HW prototype of the proposed BCC transceiver uses off-the-shelf components and has been designed to enable various parametric sweeps without changing the fundamental architecture of the transceiver.

The configurable BCC transceiver board comprises two Op Amps, eight Schmitt triggers, variable resistors, voltage biasing circuits, customizable filter circuits, Op Amp reference biasing circuits, I/O headers, and many test points. Using jumpers, all these components can be connected to, and disconnected from, each other to configure a customized transceiver. The on-board filters can be configured as low-pass, high-pass, band-pass, or band-stop filters. Additionally, the external circuits can be interfaced at any point or stage on the board through I/O headers. The configurable board is used to test a range of circuit configurations during the testing and debugging process in order to achieve a working BCC transceiver. The configurable board can also be used as an experimental kit or a breakout board for various other applications.

10.4 Testing and Verification

The experimental setup uses a set of BCC transceivers to test and verify the bi-directional body channel communication link, as shown in Fig. 10.3. There are two ends of the communication link, Node1 and Node2. At each end, the ECS1

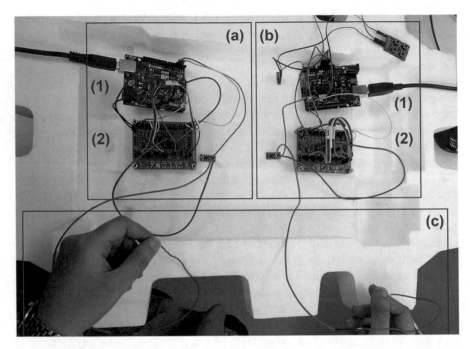

Fig. 10.3 BCC testing setup: (**a**) Node1 (**b**) Node2 (**c**) Body channel : (1) Processing unit (2) BCC transceiver

protocol (with an embedded C implementation) runs over an Arduino-101 board that is connected to the BCC transceiver. The BCC transceivers are then connected to the human body. Node1 starts communication by sending an initialization signal followed by the transmission of a 16-bit data. The data is encoded to generate a ECS1 packet, and the resulting pulses are transmitted through the human body via the BCC transceiver. The BCC transceiver at Node2 recovers the transmitted ECS1 pulses, and the ECS1 decoder on the Arduino-101 board decodes the received packet to infer the transmitted data. Similarly, Node2 transmits back the received data and Node1 receives and compares it with its own copy. In every subsequent iteration, the data is incremented, transmitted to Node2, and is compared with the received data from Node2 to validate the full bi-directional communication link. BCC communication is further verified by tapping the transmitted and received signals of the BCC link at various test points. The ECS1 signals at different stages during the reception process are shown in Fig. 10.4. Power consumption for the BCC transceiver is shown in Table 10.1 where we also compare the off-the-shelf prototype running at supply voltages of 5.5 V (Op Amp) and 3.3 V (Schmitt trigger) with a projected ASIC design running at VDD = 1.1 V. The table clearly shows that even using off-the-shelf components that are known to consume much more power, the designed BCC transceiver is at par with the power performance of the state-of-the-art ASIC BCC transceivers such as given in [25] through [42]. For each of these

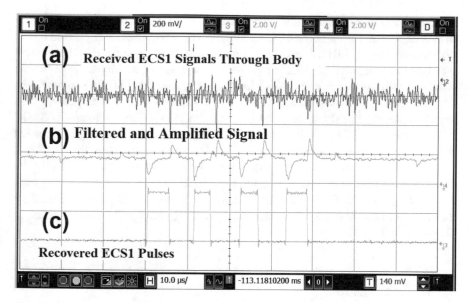

Fig. 10.4 BCC testing—oscilloscope screen-shots

Table 10.1 BCC transceiver power consumption

	5 V and 3.3 V		Scaled for 1.1 V	
	HS[a] OA[b]+ ST[c]	LS[d] OA + ST	HS OA+ ST	LS OA+ ST
Power(mW)	≈6–20	≈1–4	≈1.3–4.4	≈0.2–0.8
E_b(pJ/bit)	≈3000–31000[e]	≈10000–40000[e]	≈660–6820[e]	≈2200–8800[e]
	≈300–3100[f]	≈100–400[f]	≈70–680[f]	≈20–90[f]

[a] High speed
[b] OpAmp
[c] Schmitt trigger
[d] Low speed
[e] 50% ECS1 duty cycle
[f] 5% ECS1 duty cycle

scenarios, Table 10.1 further compares power consumption and energy efficiency
for two possible Op Amp designs: high-speed and low-speed.

As per the most recent published work, the power consumption of an Op Amp
ranges from 22 nW to 350 μW while that of a Schmitt trigger can be less than 2 μW.
Using the ECS1 protocol implemented in embedded C, the power consumption of
the BCC transceiver would be dominated by the combined power consumption
of the Op Amp and the Schmitt trigger, which ranges from 2 to 360 μW. The
custom-designed Op Amp depends on the BCC application, but its design would
be according to a competitive power budget of 100–200 μW. In the off-the-shelf
BCC transceiver, the Op Amp is low-speed with a maximum supported clock rate
of 300 KHz, resulting in a data rate in the range of 50.4–320.4 KHz with an average

of 76.8 KHz. The maximum achievable data rate of the BCC transceiver is driven by
that of ECS1 itself [57]. In the BCC context, the highest clock rate supported by the
body channel is 125 MHz. Therefore, the maximum data rate of the BCC transceiver
will be in the range of 21–133.5 Mbps with an average of 32 Mbps. The dynamic
nature of the data rate is due to the ECS1 encoding scheme [57]. In summary, the
data rate of the proposed ECS1-based BCC transceiver will be competitive with
respect to the state-of-the-art as given in [25] through [42]. The bit-error-rate (BER)
of the proposed BCC transceiver is very much dependent on the quality of the noise
filters used in the front end of the transceiver. Under the assumption that the filters
can remove all the environmental noise within the BCC bandwidth, the BER of the
ECS1-based transceiver would be zero if the separation between the falling edge
of a ECS1 pulse and the rising edge of the next PC pulse is larger than 8 ns. This
8 ns threshold is determined by measuring the settling time of the output spikes, as
described in Sect. 10.2.

10.5   Conclusions

The use of edge-coded signaling techniques such as ECS1 for BCC communication
significantly reduces the complexity of the transceiver architecture and, therefore, its
power consumption. This reduction is achieved by eliminating the need for various
complex and power-hungry circuitries, which have been the core components
of state-of-the-art BCC transceivers. The ECS1-based BCC transceivers that we
have built show successful bi-directional communication through the human body
channel by transmitting arbitrary 16-bit data words over a distance of 150*cm*
and receiving them flawlessly in a round-trip configuration. To the best of our
knowledge, this is the very first time such a BCC transmission is achieved. Future
work will tackle an integrated VLSI implementation of the ECS1-based BCC
transceiver along with the validation of such transceiver in the presence of link
non-idealities such as multipath fading, variable-ground effect, and variable skin-
electrode impedance.

Epilogue

مِنَ العلم أَنْ تعلم أَنَّك لا تعلمُ بما لا تعلم.
عبد الله بن المقفّع

The previous chapters have introduced the reader to the many facets of edge-coded signaling, including its fundamental ideas, its design principles, its hardware implementation, and its performance. The major advantages of ECS were listed in the prologue of this monograph and described in detail throughout its ten chapters. In this epilogue, we would like to summarize the research problems that are still open under the ECS paradigm. These problems include the following:

1. *ECS reliability and channel noise:* In Chap. 5, a preliminary analysis of the ECS bit error rate as a function of the signal-to-noise ratio E_b/N_0 was given and summarized in Fig. 3.3 in comparison with the BPSK digital modulation scheme. However, a more complete analysis is still needed, especially with regard to other digital modulation schemes as well as to the integration of error-correcting codes (ECC) with ECS.
2. *ECS design and non-ideal transmission channels:* It has been pointed out that the inter-symbol separation is an important ECS design parameter. The proper operation of such parameter assumes that the transmission channel remains *quiet* throughout the separation interval. One important ECS research direction is to investigate the impact of transmission channel models on ECS design and performance. Both channel bandwidth and channel noise models are important to ECS design and performance. A preliminary insight into the behavior of ECS in the presence of channel non-idealities was given in Chap. 10 where ECS was used to establish a flawless, round-trip body-coupled communication link.
3. *ECS security and lightweight cryptography:* In Chap. 7, the synergy between ECS and multilayer ciphers has been highlighted. A low-overhead, highly secure, ECS implementation has been presented based on an accelerated version of the

© Springer Nature Switzerland AG 2022
S. Muzaffar, I. M. Elfadel, *Secure, Low-Power IoT Communication Using Edge-Coded Signaling*, https://doi.org/10.1007/978-3-030-95914-2

A5/1 lightweight stream cipher. More research is needed along these directions, especially in relation to the adaptation of the many other IoT lightweight ciphers to ECS. It is possible that several of these lightweight ciphers are not amenable to the single-clock-cycle latency of the accelerated A5/1. However, we believe they are readily amenable to the multilayer crypto strengthening that is enabled by ECS.

4. *ECS hardware accelerator as a communication core:* In order to insure implementation flexibility of the ECS family of signaling protocols, we have introduced, in Chap. 8, the domain-specific, ECS RISC processor. This processor was shown to reduce embedded code complexity for ECS implementations by more than an order of magnitude. The ECS processor could serve as a dedicated communication core in the IoT sensor node. Both logical and physical IP of the ECS processor can be provided to facilitate the design of IoT nodes incorporating the ECS protocol in their communication port options along with other protocols such as UART, SPI, and I2C.

5. *ECS in the photonic domain:* Throughout this monograph, ECS hardware prototypes have been used to validate the fundamental ECS design principles. All these hardware prototypes have been electronic. Yet the basic ECS idea of coding ON bits with a pulse count can be applied to other communication domains such as the acoustic and photonic domains. In particular, the photonic domain is a promising opportunity to illustrate one of the major advantages of ECS, namely, the fact that the ECS transceiver does not need CDR circuits. For ECS to be competitive in the photonic domain, the operating frequency of the ECS transceivers must be pushed into the GHz frequency range rather than the MHz range of this monograph. Scaling up the ECS operating frequency is an important design and research challenge.

6. *Edge computing vs. communication:* As mentioned in the Prologue, our ultimate aim is to contribute to the continuing debate on edge computing vs. communication by revisiting the communication subsystem to explore the power saving opportunities that result from transceiver simplification. The power savings achieved by ECS can be exploited either to compute more or to communicate more. Either way, an ECS-based methodology to define the energy efficiency trade-off between computing and communication in an IoT node is required. The methodology should be flexible enough to determine the said trade-off on a case-by-case basis while maintaining all other ECS advantages of security, small footprint, and timing robustness.

The above ECS research directions are by no means exhaustive. Yet, we believe they represent the most promising venues under this novel IoT communication paradigm. Our hope is that this book will serve as an adequate starting point for both the experienced and budding IoT engineers who are interested in exploring these ECS research venues.

References

1. T. Akishita, H. Hiwatari, Compact Hardware Implementations of the 128-bit Blockcipher CLEFIA, in *Proceedings of Symposium on Cryptography and Information Security (SCIS 2011)*, Aug 2011, pp. 278–292
2. A. Al-Fuqaha, M. Guizani, M. Mohammadi, M. Aledhari, M. Ayyash, Internet of things: a survey on enabling technologies, protocols, and applications. IEEE Commun. Surv. Tutorials **17**(4), 2347–2376, Fourth quarter (2015)
3. R. Anderso, A5 (was: Hacking digital phones), in *Newsgroup: uk.telecom*
4. S. Banik, A. Bogdanov, T. Isobe, K. Shibutani, H. Hiwatari, T. Akishita, F. Regazzoni, Midori: a block cipher for low energy (extended version). Cryptology ePrint Archive, Report 2015/1142 (2015). https://eprint.iacr.org/2015/1142
5. E. Barkan, E. Biham, N. Keller, Instant ciphertext-only cryptanalysis of GSM encrypted communication. Technion, Technical Report CS-2006-07 (2006)
6. R. Beaulieu, D. Shors, J. Smith, S. Treatman-Clark, B. Weeks, L. Wingers, The SIMON and SPECK Families of Lightweight Block Ciphers. Cryptology ePrint Archive, Report 2013/404 (2013). https://eprint.iacr.org/2013/404
7. A. Biryukov, A. Shamir, D. Wagner, Real time cryptanalysis of A5/1 on a PC, in *Fast Software Encryption Workshop 2000*, April 2000
8. A. Bogdanov, L.R. Knudsen, G. Leander, C. Paar, A. Poschmann, M.J. Robshaw, Y. Seurin, C. Vikkelsoe, PRESENT: an ultra-lightweight block cipher, in *International Workshop on Cryptographic Hardware and Embedded Systems* (2007), pp. 450–466
9. M. Davies, N. Srinivasa, T. Lin, G. Chinya, Y. Cao, S.H. Choday, G. Dimou, P. Joshi, N. Imam, S. Jain, Y. Liao, C. Lin, A. Lines, R. Liu, D. Mathaikutty, S. McCoy, A. Paul, J. Tse, G. Venkataramanan, Y. Weng, A. Wild, Y. Yang, H. Wang, Loihi: a neuromorphic manycore processor with on-chip learning. IEEE Micro **38**(1), 82–99 (2018)
10. E. Diaconescu, C. Spirleanu, An identifying and authorizing application using 1-wire technology, in *International Symposium for Design and Technology in Electronic Packaging*, Sept 2010, pp. 243–248
11. C. dos Reis Filho, E. da Silva, E. de L. Azevedo, J. Seminario, L. Dibb, Monolithic data circuit-terminating unit (DCU) for a one-wire vehicle network, in *Proceedings of the 24th European Solid-State Circuits Conference (ESSCIRC '98), Hague*, Sept 1998, pp. 228–231
12. Q. Du, J. Zhuang, T. Kwasniewski, A 2.5 Gb/s, low power clock and data recovery circuit, in *20th Canadian Conference on Electrical and Computer Engineering (CCECE), Vancouver, BC*, April 2007, pp. 526–529
13. S. Even, O. Goldreich, On the power of cascade ciphers, in *ACM Transactions on Computer Systems (TOCS)*, May 1985, pp. 108–116

© Springer Nature Switzerland AG 2022
S. Muzaffar, I. M. Elfadel, *Secure, Low-Power IoT Communication Using Edge-Coded Signaling*, https://doi.org/10.1007/978-3-030-95914-2

14. M. Feldhofer, J. Wolkerstorfer, V. Rijmen, AES implementation on a grain of sand. IEE Proc. Inf. Secur. **152**(1), 13–20 (2005)
15. B. Fleischer et al., A scalable multi-TeraOPS deep learning processor core for AI training and inference, in *Symposium on VLSI Circuits, Honolulu*, June 2018
16. Freescale Semiconductors, (2008–2009) UICC - Contactless Front-end (CLF) Interface, Technical Specification, Version 7.3.0. https://www.etsi.org/deliver/etsi_TS/102600_102699/102613/07.03.00_60/ts_102613v070300p.pdf
17. Freescale semiconductors, MPC860 PowerQUICC Family User's Manual (2004). https://www.nxp.com/docs/en/reference-manual/MPC860UM.pdf
18. S.B. Furber, F. Galluppi, S. Temple, L.A. Plana, The SpiNNaker project. Proc. IEEE **102**(5), 652–665 (2014)
19. J.Dj. Golic, Cryptanalysis of alleged A5 stream cipher, in *International Conference on the Theory and Application of Cryptographic Techniques*, May 1997, pp. 239–255
20. T. Good, M. Benaiss, Hardware results for selected stream cipher candidates, in *State of the Art of Stream Ciphers 2007 (SASC 2007)*, Feb 2007
21. L. Han, J. Han, X. Zeng, R. Lu, J. Zhao, A programmable security processor for cryptography algorithms, in *9th International Conference on Solid-State and Integrated-Circuit Technology, Beijing*, Oct 2008, pp. 2144–2147
22. J. Hennessy, D. Paterson, A new golden age for computer architecture: domain-specific hardware/software co-design, enhanced security, open instruction sets, agile chip development. Turing Lecture at ISCA 2018 (2018). https://www.youtube.com/watch?v=3LVeEjsn8Ts
23. J.L. Hennessy, D.A. Patterson, *Computer Architecture, Fifth Edition: A Quantitative Approach*, 5th edn. (Morgan Kaufmann, San Francisco, CA, 2011)
24. D. Hong, J. Sung, S. Hong, J. Lim, S. Lee, B.S. Koo, C. Lee, D. Chang, J. Lee, K. Jeong, H. Kim, HIGHT: a new block cipher suitable for low-resource device, in *International Workshop on Cryptographic Hardware and Embedded Systems* (2006), pp. 46–59
25. J. Huang, L. Wang, D. Zhang, Y. Zhang, A low-frequency low-noise transceiver for human body channel communication, in *IEEE Biomedical Circuits and Systems Conference (BioCAS), Beijing*, Nov 2009, pp. 37–40
26. B. Huang, J. Lei, Y. Bo, The reading data error analysis of 1- wire bus digital temperature sensor DS18B20, in *International Conference on Modelling, Identification and Control*, June 2012, pp. 433–436
27. C. Jia, D. Wu, I. Hawkins, A. Forsyth, One-wire communication system for cryogenic converter control, in *6th IET International Conference on Power Electronics, Machines and Drives (PEMD 2012), Penang*, March 2012, pp. 1–5
28. N.P. Jouppi et al., In-datacenter performance analysis of a tensor processing unit, in *44th ACM/IEEE Annual International Symposium on Computer Architecture (ISCA 2017), Toronto, ON*, June 2017, pp. 1–12
29. N.P. Jouppi, C. Young, N. Patil, D. Patterson, A domain-specific architecture for deep neural networks. Commun. ACM **61**(9), 50–59 (2018)
30. S. Jun-Ren, L. Te-Wen, H. Chung-Chih, Delay-line based fast-locking all-digital pulsewidth-control circuit with programmable duty cycle, in *IEEE Asian Solid State Circuits Conference (A-SSCC)*, Nov 2012, pp. 305–308
31. M. Katagi, S. Moria, Lightweight cryptography for the internet of things, in *Sony Corporation* (2012)
32. B. Koo, D. Roh, H. Kim, Y. Jung, D.-G. Lee, D. Kwon, *CHAM: A Family of Lightweight Block Ciphers for Resource-Constrained Devices* (Springer, Berlin, 2018), pp. 3–25
33. A. Kulkarni, A. Page, N. Attaran, A. Jafari, M. Malik, H. Homayoun, T. Mohsenin, An energy-efficient programmable manycore accelerator for personalized biomedical applications. IEEE Trans. Very Large Scale Integr. VLSI Syst. **26**(1), 96–109 (2018)
34. M. Kumar, S.K. Pal, A. Panigrahi, FeW: a lightweight block cipher, in *IACR Cryptology ePrint Archive* (2014), pp. 326

35. S. Li, X. Liu, M. Mao, H.H. Li, Y. Chen, B. Li, Y. Wang, Heterogeneous systems with reconfigurable neuromorphic computing accelerators, in *2016 IEEE International Symposium on Circuits and Systems (ISCAS)*, May 2016, pp. 125–128

36. linux-mips.org, Cisco Systems Routers (2012). https://www.linux-mips.org/wiki/Cisco

37. M. Loh, A. Emami-Neyestanak, All-digital CDR for high-density, high-speed I/O, in *12th IEEE Symposium on VLSI Circuits (VLSIC'10), Honolulu, HI*, June 2010, pp. 147–148

38. M. Loh, A. Emami-Neyestanak, A 3x9 Gb/s shared, all-digital CDR for high-speed, high-density I/O. IEEE J. Solid State Circuits **47**(3), 641–651 (2012)

39. J.V. Lunteren, C. Hagleitner, T. Heil, G. Biran, U. Shvadron, K. Atasu, Designing a programmable wire-speed regular-expression matching accelerator, in *45th Annual IEEE/ACM International Symposium on Microarchitecture (MICRO-45), Vancouver, BC*, Dec 2012, pp. 461–472

40. F. Mace, F.-X. Standaert, J.-J. Quisquater, ASIC implementations of the block cipher SEA for constrained applications, in *RFID Security - RFIDsec 2007, Malaga*, July 2007, pp. 103–114

41. M. Mahmoud, A. Mohamad, A study of efficient power consumption wireless communication techniques/modules for internet of things (IoT) applications. Adv. Internet Things **6**(2), 19–29 (2016)

42. S. Maity, B. Chatterjee, G. Chang, S. Sen, A 6.3pJ/b 30Mbps-30dB SIR-tolerant broadband interference-robust human body communication transceiver using time domain signal-interference separation, in *IEEE Custom Integrated Circuits Conference (CICC)*, San Diego, CA, April 2018, pp. 1–4

43. J.L. Mauri, J.P.C. Rodrigues, Router power consumption analysis: towards green communications, in *Green Communication and Networking* (Springer, Berlin, 2013)

44. MAXIM, *OneWireViewer User's Guide, Version 1.4* (2009)

45. T. Mehrabi, K. Raahemifar, V. Geurkov, Design of a 4-bit programmable delay with TDC-based BIST for use in serial data links, in *International Symposium on Integrated Circuits (ISIC)*, Dec 2014, pp. 580–583

46. A. Menon, S. Murugan, C. Rebeiro, N. Gala, K. Veezhinathan, Shakti-T: a RISC-V processor with light weight security extensions, in *Proceedings of the Hardware and Architectural Support for Security and Privacy*, ser. HASP '17 (ACM, New York, 2017), pp. 2:1–2:8

47. B.J. Mohd, T. Hayajneh, Lightweight block ciphers for IoT: energy optimization and survivability techniques. IEEE Access **6**, 35966–35978 (2018)

48. S. Muzaffar, I.M. Elfadel, Power management of pulsed-index communication protocols, in *33rd IEEE International Conference on Computer Design (ICCD), New York, NY*, Oct 2015, pp. 375–378

49. S. Muzaffar, I.M. Elfadel, Timing and robustness analysis of pulsed-index protocols for single-channel IoT communications, in *23rd IFIP/IEEE International Conference on Very Large Scale Integration (VLSI-SoC 2015), Daejeon*, Oct 2015, pp. 225–230

50. S. Muzaffar, I.M. Elfadel, A versatile hardware platform for the development and characterization of IoT sensor networks, in *59th IEEE International Midwest Symposium on Circuits and Systems (MWSCAS'16), Abu Dhabi*, Oct 2016, pp. 1–4

51. S. Muzaffar, I.M. Elfadel, A pulsed decimal technique for single-channel, dynamic signaling for IoT applications, in *25th IFIP/IEEE International Conference on Very Large Scale Integration (VLSI-SoC 2017), Abu Dhabi*, Oct 2017, pp. 1–6

52. S. Muzaffar, I.M. Elfadel, An instruction set architecture for low-power, dynamic IoT communication, in *26th IFIP/IEEE International Conference on Very Large Scale Integration (VLSI-SoC 2018), Verona*, Oct 2018 (Accepted)

53. S. Muzaffar, I.M. Elfadel, A domain-specific processor microarchitecture for energy-efficient, dynamic IoT communication. IEEE Trans. Very Large Scale Integr. VLSI Syst. **27**(9), 2074–2087 (2019)

54. S. Muzaffar, I.M. Elfadel, A self-synchronizing, low-power, low-complexity transceiver for body-coupled communication, in *41st Annual International Conference of the IEEE Engineering in Medicine and Biology Society (EMBC 2019), Berlin* (2019)

55. S. Muzaffar, I.M. Elfadel, Double data rate dynamic edge-coded signaling for low-power IoT communication, in *27th IFIP/IEEE International Conference on Very Large Scale Integration (VLSI-SoC 2019), Cuzco*, Oct 2019, pp. 317–322

56. S. Muzaffar, I.M. Elfadel, IoT communication using dynamic edge-coded serial signaling. ACM Trans. Sen. Netw. **17**(1), Article 8, 24 pp. (2021)

57. S. Muzaffar, A. Shabra, J. Yoo, I. M. Elfadel, A pulsed-index technique for single-channel, low power, dynamic signaling, in *Design, Automation and Test in Europe (DATE'15), Grenoble*, March 2015, pp. 1485–1490

58. S. Muzaffar, N. Saeed, I.M. Elfadel, Automatic protocol configuration in single-channel low-power dynamic signaling for IoT devices, in *24th IFIP/IEEE International Conference on Very Large Scale Integration (VLSI-SoC 2016), Tallinn*, Sept 2016, pp. 1–6

59. S. Muzaffar, O.T. Waheed, Z. Aung, I.M. Elfadel, Single-clock-cycle, multilayer encryption algorithm for single-channel IoT communications, in *IEEE Conference on Dependable and Secure Computing (DSC 2017), Taipei*, Aug 2017, pp. 153–158

60. OpenCores.org, openMSP430 (2018). http://opencores.org/project,openmsp430

61. OpenCores.org, openMSP430 :: Software development tools (2018). https://opencores.org/project/openmsp430/software%20development%20tools

62. D. Patterson, 50 years of computer architecture: from the mainframe CPU to the domain-specific TPU and the open RISC-V instruction set, in *IEEE International Solid - State Circuits Conference (ISSCC 2018), San Francisco, CA*, Feb 2018, pp. 27–31

63. A. Poschmann, Lightweight cryptography - cryptographic engineering for a pervasive world, in *IACR ePrint archive 2009/516* (2009)

64. Postscapes, IoT Standards and Protocols. https://www.postscapes.com/internet-of-things- protocols/

65. K.V.K.K. Prasad, *Principles of Digital Communication System and Computer Network* (Dreamtech Press, New Delhi, 2003)

66. J. Proakis, M. Salehi, *Digital Communications*, 5th edn. (McGraw-Hill Education, New York, 2008)

67. B. Razavi, *Design of Integrated Circuits for Optical Communications*, 2nd edn. (Wiley, New York, 2012)

68. W. Saadeh, Y. Yonatan, J. Yoo, A hybrid OFDM body coupled communication transceiver for binaural hearing aids in 65nm CMOS, in *IEEE International Symposium on Circuits and Systems (ISCAS), Lisbon*, May 2015, pp. 2620–2623

69. W. Saadeh, M.A.B. Altaf, H. Alsuradi, J. Yoo, A 1.1-mW ground effect-resilient body-coupled communication transceiver with pseudo OFDM for head and body area network. IEEE J. Solid State Circuits **52**(10), 2690–2702 (2017)

70. J. Sawada, F. Akopyan, A.S. Cassidy, B. Taba, et. al., TrueNorth ecosystem for brain-inspired computing: scalable systems, software, and applications, in *SC '16: Proceedings of the International Conference for High Performance Computing, Networking, Storage and Analysis*, Nov 2016, pp. 130–141

71. K. Shibutani, T. Isobe, H. Hiwatari, A. Mitsuda, T. Akishita, T. Shirai, Piccolo: an ultra-lightweight blockcipher, in *International Workshop on Cryptographic Hardware and Embedded Systems* (2011), pp. 342–357

72. T. Shirai et al., The 128-bit blockcipher CLEFIA, in *International Workshop on Fast Software Encryption* (2007)

73. L.-K. Soh, W.-T. Wong, A 2.5-12.5 Gbps interpolator-based clock and data recovery circuit for FPGA, in *4th Asia Symposium on Quality Electronic Design (ASQED), Penang*, July 2012, pp. 373–379

74. M. Stamp, *Information Security: Principles and Practices*, 2nd edn. (Wiley, New York, 2011)

75. V. Sze, Y. Chen, T. Yang, J.S. Emer, Efficient processing of deep neural networks: a tutorial and survey. Proc. IEEE **105**(12), 2295–2329 (2017)

76. R. Teja, B.R. Jammu, M. Adimulam, M. Ayi, VLSI implementation of LTSSM, in *International Conference of Electronics, Communication and Aerospace Technology (ICECA 2017), Coimbatore*, April 2017, pp. 129–134

77. Y. Urano, W.-J. Yun, T. Kuroda, H. Ishikuro, A 1.26mW/Gbps 8 locking cycles versatile all-digital CDR with TDC combined DLL, in *45th IEEE International Symposium on Circuits and Systems (ISCAS'13), Beijing*, May 2013, pp. 1576–1579
78. S. Wang, K.W.E. Cheng, K. Ding, Design of the temperature and humidity instrument based on 1-wire sensor for electric vehicle motors, in *International Conference on Power Electronics Systems and Applications*, May 2009, pp. 1–5
79. A.X. Widmer, P.A. Franaszek, A DC-balanced, partitioned-block, 8B/10B transmission code. IBM J. Res. Dev. **27**(5), 440–451 (1983)
80. W. Wu, L. Zhang, LBlock: a lightweight block cipher, in *International Conference on Applied Cryptography and Network Security* (2011), pp. 327–344

Index

© Springer Nature Switzerland AG 2022
S. Muzaffar, I. M. Elfadel, *Secure, Low-Power IoT Communication Using Edge-Coded Signaling*, https://doi.org/10.1007/978-3-030-95914-2

Printed in the United States
by Baker & Taylor Publisher Services